Charles Johnson Maynard

The Naturalist's Guide

In Collecting and Preserving Objects of Natural History

Charles Johnson Maynard

The Naturalist's Guide
In Collecting and Preserving Objects of Natural History

ISBN/EAN: 9783743408258

Manufactured in Europe, USA, Canada, Australia, Japa

Cover: Foto ©berggeist007 / pixelio.de

Manufactured and distributed by brebook publishing software
(www.brebook.com)

Charles Johnson Maynard

The Naturalist's Guide

THE

NATURALIST'S GUIDE

IN

COLLECTING AND PRESERVING OBJECTS
OF NATURAL HISTORY.

BY

C. J. MAYNARD.

Illustrated.

BOSTON:
CUPPLES AND HURD, PUBLISHERS,
THE ALGONQUIN PRESS.
NEWTONVILLE
C. J. MAYNARD AND COMPANY.

2579

Are You Interested
IN
Natural History?

THEN you want to send ten cents for our new Catalogue (with over 125 illustrations) of Tools and Supplies for Naturalists. We carry in stock or can supply at short notice any Tools or Supplies used by Naturalists.

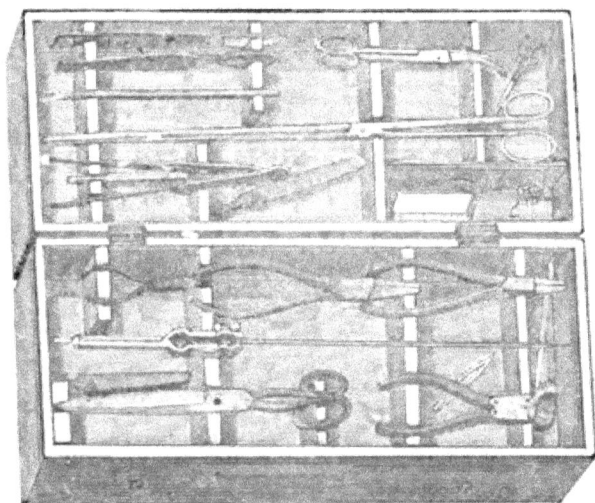

CHAMPION TOOL BOX.

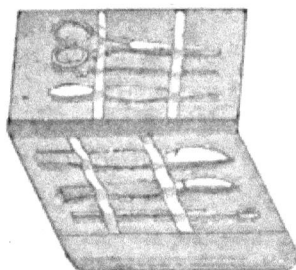

TAXIDERMIST POCKET-CASE.

SCALPELS.
Ebony handle, large, medium, small. All steel, medium.

CARTILAGE KNIFE.

Extra heavy.   Also the Hornaday Skinning Knife.

SURGEON'S DISSECTING SCISSORS.

GLASS EYES.

If you Mount Birds you will need GLASS EYES, and we would ask that you examine the HURST EYE, which we claim to be THE BEST in the market, not only for the natural effect which they give but for their freedom from cracks and flaws, and you will find them *as large size as is shown in our diagrams.*

We can also supply Leaves, Grasses, Mosses, Sands and Smalts for Cases, Trees and Backgrounds in innumerable variety.

STUMPS.

Natural, on bases decorated.
Finished Rock-work or Stands for Birds or Animals, Shields and Panels for Heads and Game Pieces.

SQUARE GLASS SHADES.

With oval, convex centres, for wall pieces.   With or without Frames, and Painted on Plush Backs.

10 x 14, for Quail, Snipe, Woodcock, etc.
16 x 22, for medium-sized Ducks, Grouse, etc.
18 x 26, large size and pairs of Ducks, etc.
22 x 38, for large-spread Birds, Gulls, Hawks, Owls.

Entomologists, Botanists and Oologists.

Will find what they want, as well as the Ornithologist.

COLLECTION BOX.

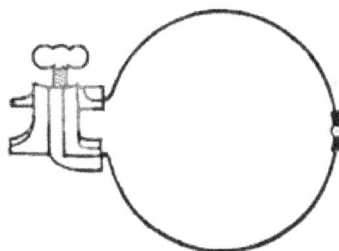

COLLECTING NETS.

COLLECTING CANS.

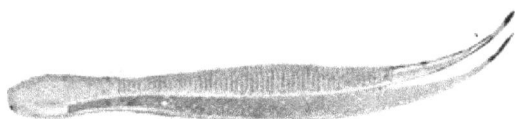

FORCEPS.

EGG DRILLS.

Best style, extra fine steel, No. 1, 3-32 inch burr, short or long handle.
"   "   "   "   "   No. 2, 5-32 inch burr   "   "   "
"   "   "   "   "   No. 3, 6-32 inch burr   "   "   "
"   "   "   "   "   No. 4, 8-32 inch burr   "   "   "
"   "   "   "   "   No. 5, 12-32 inch burr, long handle.
"   "   "   "   "   No. 6, 16-32 inch burr   "   "

Egg Blowers, Insect Pins, Drying and Mounting Paper, Data Blanks, etc.

BOOKS

On every branch of Natural History.

TAN YOUR SKINS

WITH

·TANNINE.·

Every sportsman, taxidermist and naturalist should have on hand a bottle of this Liquor. With it a skin can be tanned quickly and without trouble.

Every sportsman has experienced the sensation of the most poignant disappointment to arrive home and find that the hair is all coming off the Fox, Bear or Deer skin that he has salted and packed so carefully for many a long mile. It might have been avoided if he had a bottle of *Tannine*.

We have testimonials from every section; the best one is to see the work it does. It speaks for itself, and talks right out aloud.

PRICE, $1.00,
With Full Directions for Use.

We are establishing agencies in various sections of the country, in order to place it within easy reach of the consumer.

GIVE · IT · A · TRIAL.

Frank Blake Webster Company,
NATURALISTS' SUPPLY DEPOT,
409 Washington Street, Boston, Mass.

INTRODUCTION.

THE great need of a good illustrated work to guide young naturalists in collecting and preserving objects of natural history has induced me to prepare the present Manual. In this attempt I hope I have been in some degree successful. I have spared no pains to bring together, in a comprehensive form, the results of many years of experience in collecting and preserving objects of natural history, both for private cabinets and for scientific museums.

No popular work of this kind has before been published in America. Throughout the present work I have endeavored to encourage the young to engage in the ennobling study of Natural History, and to join the band of young naturalists so rapidly increasing in our land.

I trust the reader will not by any means keep the teachings of this book secret, as some taxidermists are wont to counsel, but spread it broadcast among those who would profit by the information I have herein attempted to convey. It is intended for the NATURALIST, whoever and wherever he may be ; and as it comes from a colaborer in the common field, it will, perhaps, be well received.

All of Part First is original. In preparing objects of natural history I have in a great degree invented methods of my own, and have not given in this work a single one that I have not tested and proved equal to all others, if not superior. To avoid confusion, I have given only the method which experience has taught me to be the best.

In this connection my thanks are due to Mr. E. L. Weeks, whose excellent illustrations will be found to add greatly to the value of the work.

CONTENTS.

———◆———

CHAPTER I.

COLLECTING AND PRESERVING BIRDS.

CHAPTER II.

COLLECTING AND PRESERVING MAMMALS.

CHAPTER III.

COLLECTING AND PRESERVING INSECTS FOR THE CABINET.

CHAPTER IV.

COLLECTING AND PRESERVING FISHES AND REPTILES.

CHAPTER V.

MISCELLANEOUS COLLECTIONS.

CHAPTER VI.

LIST AND EXPLANATION OF PLATES.

PLATE I. INSTRUMENTS used in preparing birds, etc., and for blowing eggs. *Fig.* 1, Common Pliers; *Fig.* 2, Cutting Pliers; *Fig.* 3, Tweezers; *Fig.* 4, Scalpel; *Figs.* 5 and 6, Egg-drills; *Fig.* 7, Blow-pipe; *Fig.* 8, Hook for removing embryos from eggs.

PLATE II. — WINGS, showing the positions of the different feathers, as follows: —

Fig. 1. *Wing of a Red-tailed Hawk* (*Buteo borealis*, Vieill.). — a indicates the primaries, or quills; b, secondaries; c, tertiaries; d, scapularies; g, greater wing-coverts; f, lesser wing-coverts; e, spurious wing, or quills.

Fig. 2. *Wing of a Coot, or Mud Hen* (*Fulica Americana*, Gmelin). — a indicates the primaries, or quills; b, secondaries; c, tertiaries; d, scapularies; e, spurious wing, or quills.

The tertiaries and scapularies are elongated in most of the aquatic birds, and in some of the Waders. They are *always prominent*, if not elongated, on long-winged birds, such as the Eagles, Hawks, Owls, Vultures, etc.; while they are only rudimentary on short-winged birds, such as the Thrushes, Warblers, Sparrows, etc.

PLATE III. HEAD OF THE BALD EAGLE (*Haliætus leucocephalus*, Savigny), showing the different parts, as follows: — a, the throat; b, chin; c, commissure, or the folding edges of the mandibles; d, under mandible; s, gonys; p, gape; g, upper mandible; h, culmen; i, tip; j, base of bill; k, cere (naked skin at the base of the upper mandible, prominent in the rapacious birds); l, frontal feathers; m, lores; n, crown; o, occiput.

The irides are the colored circles that surround the pupil. The color of these decides the so-called "color of the eye."

PLATE IV. ILLUSTRATES PREPARING SKINS. — *Figs.* 1 and 2. Corrugated board, used in drying skins; d, skin on the board, in the proper position. *Fig.* 3. A "skin" prepared for scientific use; ♂, label, on which is marked the number and sex.

PLATE V. DISSECTED SONG SPARROW (*Melospiza melodia*, Baird), illustrating the sexes in the breeding season, as follows : — *Fig.* 1. An adult female (♀); 1,1, peculiar yellow glands; 2, ovary; 3, oviduct; 4, lungs. *Fig.* 2. An adult male (♂); 1, lungs; 2, peculiar yellow glands; 3, 3, testicles.

PLATE VI. DISSECTED SONG SPARROW, illustrating the sexes of the young-of-the-year, in autumn, as follows : — *Fig.* 1. A young male (♂); 1, lungs; 2, 2, yellow glands; 3, 3, testicles. *Fig.* 2. A young female (♀); 1, 1, yellow glands; 2, ovary; 3, lungs; 4, oviduct.

PLATE VII. OUTLINE OF GROUSE, showing the position of the different parts, as follows : — a, the back; b, rump; c, upper tail-coverts; d, under tail-coverts; e, vent; f, tibia; g, tarsi; h, breast; i, side; j, neck; k, hind neck; l, abdomen; m, feet; n, throat.

PLATE VIII. ILLUSTRATES MOUNTING BIRDS. — *Figs.* 1, 2. Artificial body; a, bone of leg; b, wire bent; c, wire clenched; f, h, tail wire. *Fig.* 3. Mounted bird; a, perpendicular line, showing the position of the head compared with the feet and base of the stand; b, b, wires for retaining the upper part of the wing in position; c, c. wires for retaining the lower part of the wing in position; e, e, wires for the tail; d, showing the tail-feathers plaited; f, stand. *Fig.* 4. Stand for mounting birds with the wings extended; b, b, parallel wires; c, wires bent; a, block of wood for the bottom of the stand, *Fig.* 5. Head of Cedar-Bird, to illustrate the elevating of the crest; g, cotton on the pin; b, feathers of the crest in position on the cotton.

PLATE IX. ILLUSTRATES MOUNTING MAMMALS. — *Fig.* 1. A, plank for supporting iron rods; 8, iron rod for supporting head; 14, cap, nut, and screw for fastening the end of the rod in the skull; 7, 7, 7, 7, iron rods to support the body; 5, 6, 5, 6, caps, etc. for fastening the upper ends of the rods to the plank; 17, 17, 17, 17, caps, etc. for fastening the lower part of the rods to the stand (10); 15, wire for supporting the tail; 16, 16, 16, 16, 16, 16, 16, 16, artificial sections of

hemp, grass, or plaster used as a substitute for the natural body.
Fig. 2. A, nut; B, cap; C, thread.

PLATE X. SKELETON OF A GROUSE, OR PRAIRIE HEN (*Cupidonia cupido*, Baird), showing the different bones, as follows:—
a, the skull; b, vertebra of the neck; c, humerus; d, forearm; f, phalanges; g, furcula; h, sternum; i, marginal indentations; j, thigh; k, tarsus; y, tibia; m, rump; n, coccygus; A, ribs; B, lower joint of thigh.

DIRECTIONS

FOR COLLECTING, PRESERVING, AND MOUNTING

BIRDS, MAMMALS, FISHES,

ETC., ETC., ETC.

THE NATURALIST'S GUIDE.

CHAPTER I.

COLLECTING AND PRESERVING BIRDS.

SECTION I. *How to collect.* — Personal experience is a good, and in fact the only adequate, teacher we can have in learning any art. The need of such a teacher is felt by none more than by the naturalist who wishes to bring together a complete collection of the birds of even his own immediate district. Hence I trust I shall not be accused of egotism, if, in this section, I endeavor to impart to the reader some things that experience has taught me.

It is of first importance for the collector to gain as complete a knowledge as possible of the notes and habits of birds, and of the localities frequented by those he wishes to procure. This knowledge may be gained by carefully studying the writings of men who have paid particular attention to the subject. *Too* much dependence must not be placed on books, as the best of these contain error as well as truth ; besides, birds are very variable in their habits in different localities. The collector must then depend mainly upon himself. He must visit *every* locality, — the mountain-top and the dark swampy thicket, as well as the meadow, the plain, or the open forest, as in each of these localities he will find species that he may not meet elsewhere. A little patience will help any one through the worst of places.

The quaking bog, where a misstep may plunge the adventurer into the slimy ooze, is also an excellent locality for

certain species. But when the collector returns home wet
and hungry, fatigued and disheartened, — as he now and
then will, — let him not be discouraged. Try again! the
next day, and even the next, if need be, until the desired
specimen is obtained. After all, the earnest naturalist will
be amply rewarded for the exercise of patience and perseverance by securing a rare specimen.

The *true* naturalist never thinks of cold and disappointment, of days of fatigue and hours of patient watching,
when at last he holds in his hand the long-searched-for
bird. Ample reward is this for all his former trials ; he is
now ready to go into bog and through brier. And thus the
enthusiastic naturalist travels on, not discouraged by toil
and trouble, laughed to scorn by the so-called "practical"
men, who are unable to appreciate his high motive. This,
however, he forgets when in field or study he meets with
the cordial greeting of his brother naturalist, as they with
mutual interest relate their discoveries and adventures.

To the travelling collector a few special hints are necessary. While visiting a remote region, but little known,
one should not neglect to shoot numbers of every bird met
with, even if they are common species at home, as they
will not only furnish data on the distribution of the species,
but they may present interesting characters peculiar to
·that locality. If a certain species appears common, do not
delay collecting specimens, for peculiar circumstances may
have brought them together in unusual numbers ; at some
future time they may be rare.

A well-trained dog is of great value while collecting
birds, especially the Quails, Marsh Wrens, Sea-side and
Sharp-tailed Finches, — in fact, all birds 'that are difficult to
start in open meadows and grassy places. While searching thickets, great watchfulness should be observed, especially in the autumn, when many birds have no conspicuous
note, otherwise many of the more wary of the Warblers

will escape notice. The slightest chirp should be carefully followed; the slightest motion of the branches closely watched. If a bird is seen that is not fully recognized, it should be shot at once, for in no other way can it be determined whether it is not a *rara avis*.

By carefully watching the motions of birds, the collector will soon become so expert as to be able generally to distinguish the different species of Warblers, even at a distance. Carefully scrutinize also the tops of tall forest-trees, as I have there taken, in autumn, some of the rarest Warblers.

In spring male birds are quite readily found, as they are then in full song; but the same caution must be used in collecting females that is practised in autumn, as they are generally shy and difficult to find. Hence it is a good rule always to secure the female *first*, when she is seen with the male; for, in spite of all the collector's efforts, he will find that there will be four males to one female in his collection.

During winter some birds may be found in the thick woods that one would hardly expect to find at this season, such as the Robin, Golden-winged Woodpecker, etc. The open fields should not be neglected even during snow-storms, as it is then that such ordinarily cautious birds as the Snowy Owl may be approached quite readily; or the capture of a Jerfalcon may reward the collector for a disagreeable tramp. The salt marshes and sandy sea-shores are the resort of a great many winter birds, and the collector will perhaps find himself amply repaid for a few visits to these localities at this season.

Do not neglect to collect the young of birds; by procuring specimens of these from the time they become fully fledged until they attain the perfectly mature plumage, one becomes familiar with all the stages through which a given species passes, and will thus avoid many errors into which some of our eminent ornithologists have fallen, —

that of mistaking the young of certain well-known birds for a different species from the adult, from not being acquainted with the immature stages. All birds should be taken that exhibit any unusual characters, such as unusually large or small bills or feet; or change of plumage, such as very pale, or very bright, cases of albinism, etc.

The gun used by a collector should have a small bore, not larger than No. 14, for shooting small birds; for Ducks, and other large water-birds, one of larger calibre will be found more effectual. The best shot to use for small birds is "Dust shot," if it can be procured; if not, No. 12 will answer. No. 8 will do for Ducks and large birds. For Hawks and Eagles, Ealy's wire cartridges are the best.

In shooting small birds, load as lightly as possible. Put in no more shot than is required to kill the bird. As you can approach very near most small birds, you will find, by experiment, that you can kill them with very little shot. If too much powder is used, it will impel the shot with so much force as to send it completely through the bird, thereby making *two* holes, when less powder, by causing less force, would have made only *one*, and the bird would have been killed just as effectually. When shot goes into the body of a bird, it generally carries feathers with it, and in a measure plugs the hole; but when it is forced through and comes out, it often carries away a small patch of feathers and skin, leaving an open wound, from which the blood flows freely.

If the bird is not instantly killed by shooting, the thumb and forefinger should be placed with a firm pressure on each side of its body under the wings, when it will soon die. This operation compresses the lungs and prevents the bird's breathing. Besides mercifully ending its suffering, its death causes the flow of blood in a great measure to cease, for this reason it should be killed as quickly as possible.

The mouth, nostrils, and vent should now be plugged with cotton or tow. By blowing aside the feathers the shot-holes may be detected; if they bleed, or are in the abdomen or rump, a pinch of calcined plaster * should be placed upon them; this absorbs the blood, or any fluid that may ooze out. When shot enters either the abdomen or rump, it is apt to cut the intestines and set free the fluids contained therein. If the blood has already soiled the feathers, remove as much as is possible with a knife, then sprinkle plaster on the spot, and rub the soiled feathers gently between the thumb and fingers; this, if repeated, will generally remove any spots of blood, etc., if the operation is performed before the blood becomes dry. When the blood is dry, it is removed after the bird is skinned, as will be hereafter described.

Next make a note of the color of the eyes, feet, and bill of the specimens, also note the color of the cere in birds of prey, and the naked skin of the lores and about the bill of the Herons, also about the heads of the Vultures. After smoothing the feathers carefully, place the bird in a paper cone,† head first, then pin or twist up the larger end, taking care not to injure the tail-feathers. The blood can be washed from the feathers of all the swimmers, but the bird, in this case, should be allowed to dry before packing in paper. If grease or oily matter has oozed out upon the feathers, the bird should not be washed, but the plaster be used as before, only in larger quantities.

All traces of blood should be instantly removed from white feathers, as it is very apt to stain them if it remains upon them long. The paper containing the bird should

* This is burned plaster or gypsum, and is used by stucco-makers. If it cannot be procured, the unburned plaster or common ground gypsum used by farmers, or air-slacked lime, pulverized chalk, or ashes, — in fact, anything that will absorb the blood, — will answer.

† The leaves of an old pamphlet are about the right size for making cones for small birds, and can be easily obtained.

be placed in a light basket, — a willow fish-basket is the
best for this purpose, — suspended by a strap over the shoulder, and resting upon the hip.   If there are but one or two
birds in the basket, it should be filled with grass, or loose
paper, to keep them steady, as otherwise they might receive injury by rolling from side to side.   In packing birds,
avoid putting the largest at the top, as their weight will
cause the smallest to bleed.   Do not hold a bird in the
hand any longer than is necessary ; if possible, take it by
the feet or bill, for the perspiration from the hand tends to
impair the gloss of the plumage.

A good collector must practise, in order to become a good
shot.   He must always keep his gun in readiness, for at
any moment a bird that he desires may start up at his feet,
or peer out from the bushes for only an instant before
flying away; by being ready, he will thus secure many
birds that he would otherwise lose.

To be in readiness at all times, the gun should be carried in the hollow of the left arm, with the muzzle pointed
backwards, or with the stock under the right arm, with the
muzzle pointed towards the ground, which is undoubtedly
the safest way, especially if you are hunting with a com-
panion.   Too much caution cannot be used in handling a
loaded gun, especially by a professional collector, who may
spend two thirds of his time with a gun in his hand.   A
gun should never be carried in other than three ways, —
the two above mentioned and directly over the shoulder.
If the collector becomes accustomed to these ways, which
are all perfectly safe, he will never think of any other.
Surely, this caution is necessary to one who is travelling
through all sorts of places, when a slip or a fall with a care-
lessly held gun might cripple him for life, by an accidental
discharge.

While passing through thick bushes, *always* carry the
gun under the arm, as this prevents its accidental dis-

charge by the bushes catching the trigger or hammer.
Never allow the muzzle of the gun to point at any one,
even for an instant. All these things depend upon habit,
and will cause a thoughtful man, who has handled a gun for
a long time, to be much more careful than a person who
seldom takes one in his hands. The thoughtful man
prefers rather to avoid accidents to himself and others —
by care in advance — than to risk the chance of having to
mourn his carelessness afterwards. The various devices for
snaring birds are undoubtedly the best ways to secure them
without injuring their plumage. But the collector will
have to rely mainly upon his gun ; and by following the
above instruction regarding the light charges, he will find
that he will generally kill a bird without injuring its
plumage seriously. If he carefully attends to it afterward
in the way described, he will save himself much trouble
when he wishes to preserve it.

In an old French cook-book may be found a receipt for a
rabbit-stew, commencing with, "First, catch your rabbit,"
etc., — which rule is applicable to the collector. First, study
with attention the art of collecting. Many and long have
been the lessons in collecting that I have taken in long tramps
through sunshine and storm, in the bracing air among the
mountains of Northern Maine and New Hampshire, on
sandy islands and rocky shores, amid the luxuriant forests
and along the rivers and lagoons of semi-tropical Florida.
Hours of danger and perplexity have been mingled with
days of inexpressible pleasure, which all must experience
who study from the Great Book of Nature. Not easily,
then, I may add, have I learned what I am trying to impart to others in these pages.

Since writing the preceding, I have been informed by
my friend, Mr. W. Brewster, of Cambridge, that in collecting such small birds as the Warblers, Sparrows, Wrens, etc.,
he has used a "blow-gun" to great advantage, constructed
1*

somewhat after the pattern of the celebrated instrument
that is used by the natives of some portions of South
America to shoot poisoned arrows. His "gun" is made
of pine-wood, and is about four feet and a half long; it is
bored smoothly the whole length with a quarter-inch hole.
For ammunition Mr. Brewster uses balls made of soft
putty. These, blown at birds, will hit them hard enough
to kill, if the gun be aimed rightly, which art can be
acquired by practice. This is certainly the preferable way
to collect small birds, as it does the plumage no harm. I
would suggest, however, that a tube of thin brass be used
in place of wood; if it were longer, say six feet, it would carry
with greater force and more accuracy. Glass would be still
better, if it could be supported by wood to prevent breakage, as it would be much smoother. The balls of putty
should be made to fit moderately tight. I have never
tried this method myself, but Mr. Brewster has, in a satis-
factory manner, as described above. I only wait an oppor-
tunity to test them myself, and trust that others will do
the same.

SECTION II. *How to prepare Specimens. Instruments,
Materials, etc.** — The instruments needed in preserving
birds and mammals are : a pair of common pliers, Plate I.
Fig. 1 ; a pair of cutting pliers, Fig. 2 ; a pair of tweezers,
Fig. 3 ; a scalpel, Fig. 4 ; two brushes,—one soft, the other
stiff ; a flat file, and needles and thread.

The materials needed are : wire of annealed iron of sizes
between 26 and 10, also some very fine copper wire ;
common thread, coarse and fine, also some very fine, soft
thread from the cotton-factories, — this is wound on what
are called "bobbins"; it is used in the manufacture of
cloth,—cotton tow or hemp, and fine grass ; for the latter
the long tough kind that grows in the woods is the best.

---

* All the instruments and the wire may be procured at the hardware
stores in the cities or larger towns.

Fig. 8

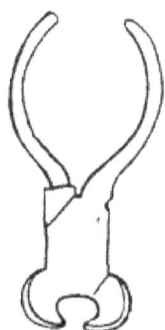

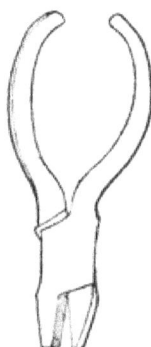

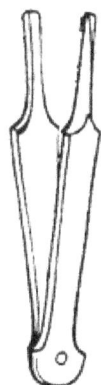

Fig. 2

Fig.1

Fig. 3

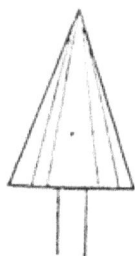

Fig. 5

Fig.7

Fig. 4

Fig.6

Plate I.

Arsenic is the best substance that can be used in preserving skins, and the only one necessary. Other preparations are no better, and often much *worse*. Strange as it may appear to some, I would say avoid especially all the so-called arsenical soaps; they are at best but filthy preparations; beside, it is a fact to which I can bear painful testimony, that they are — especially when applied to a greasy-skin — poisonous in the extreme. I have been so badly poisoned, while working upon the skins of some fat water-birds that had been preserved with arsenical soap, as to be made seriously ill, the poison having worked into the system through some small wounds or scratches on my hands. Had pure arsenic been used in preparing the skins the effect would not have been as *bad*, although grease and arsenic are generally a blood poison in *some* degree; but when combined with "soap," the effect — at least, as far as my experience goes — is much more injurious.

Arsenic alone will *sometimes* poison *slightly* the wound with which it comes in contact, but no more than common salt. There will be a slight festering and nothing more; but, on the contrary, when combined with fat, a poison is generated that must be carefully guarded against. It sometimes works under the nails of the fingers and thumbs, while one is at work skinning (especially if the birds are fat). Rubber cots should be put upon the fingers or thumbs the instant the slightest wound is detected, whereby much pain may be avoided at a small cost. The cots alluded to can be procured of almost any druggist for ten cents each.

Arsenic, however, cannot be used with too great care, as it is a deadly poison. In no case should it be left in the way of children. I have a drawer, wide, long, and shallow, in the bench at which I work upon birds, where my arsenic is kept safely, and it is always accessible. But there is probably not so much danger attending the use

of pure dry arsenic as people generally suppose. I have
been told repeatedly, by competent physicians, that the
small quantity taken, either by inhalation while using it,
or by numerous other accidental ways, would be beneficial,
rather than injurious; but be that as it may, I have used
dry arsenic constantly for ten years, and have not yet, I
think, experienced any injurious effects from it. It must
be remembered that I have, of course, used it carefully.
When used with care, in the ordinary manner, it is undoubtedly the safest and the best material that can be
used in preparing skins for the cabinet. I have never yet
had a skin decay, or attacked by moths, that was well preserved by the use of arsenic. Arsenic is very cheap, varying from five to ten cents per pound by the wholesale, and
retailed at twenty-five cents by druggists, but when bought
by the ounce the price is enormous.

There is, however, another poison to which one is exposed
while skinning animals, which cannot be too carefully
guarded against, for it is much more injurious in its effects
than fat and arsenic. I speak of the animal poison that
results from the first stage of decomposition. If on a warm
day one skins birds from which an offensive odor arises,
and a peculiar livid or purplish appearance of the skin
upon the abdomen is seen, and the intestines are distended
with an extremely poisonous gas, — which is the source
of the offensive, sickening odor,— there is danger of being poisoned. When this gas is inhaled, or penetrates the
skin through the pores (which are generally open on a
warm day), a powerful and highly dangerous poison is apt
to be the result.

In a few days numerous pimples, which are exceedingly
painful, appear upon the skin of the face and other parts
of the person, and upon those parts where there is a chaf-
ing or rubbing become large and deep sores. There is a
general languor, and, if badly poisoned, complete prostration

results; the slightest scratch upon the skin becomes a festering sore. Once poisoned in this manner (and I speak from experience), one is never afterwards able to skin any animal that has become in the *least* putrid, without experiencing some of the symptoms above described. Even birds that you handled before with impunity, you cannot now skin without great care.

The best remedy in this case is, as the Hibernian would say, not to get poisoned, — to avoid skinning all birds that exhibit the slightest signs of putrescence ; this is especially to be guarded against in warm weather, and in hot climates, where I have seen a single hour's work upon putrid birds nearly prove fatal to the careless individual.

If you *get* poisoned, bathe the parts frequently in cold water ; and if chafed, sprinkle the parts, after bathing, with wheat flour. These remedies, if persisted in, will effect a cure, if not too bad ; then, medical advice should be procured without delay.

It is just as easy to skin fresh birds as putrid ones, and much pleasanter, and in this way the evil will be avoided. If it is necessary to skin a putrid bird, — as in the case of a rare specimen, — a good bath of the hands and face in clear, cold water will entirely prevent the poison from taking effect, provided the skinning is not protracted too long. But generally, if the bird is putrid, I would advise the collector to throw it away, and obtain others that are safer to skin.

If birds and mammals are injected, by means of a small glass syringe, with a small quantity of carbolic acid at the mouth and vent, it will prevent decomposition from taking place immediately. After injecting, the mouth and vent should be plugged to prevent the acid from staining the feathers. Birds injected in this way for three successive days will continue fresh for a long time, and, if kept in a dry place, will harden completely without decomposing.

They may afterwards be skinned, **as** will be described hereafter. Impure carbolic acid will **answer as** well as the refined, and it is **much** cheaper.

The cost of this acid **is** trifling, and it will often **prove** beneficial in preserving birds in warm weather when they cannot be skinned immediately. But I would not advise its use in preserving birds when it can possibly be avoided, as it dulls the plumage, **and is offensive in** its odor in connection with **the juices of the birds** while they are being skinned. It is, perhaps, needless to **add that** this acid is a dangerous internal poison; it also burns the skin badly **when** allowed **to come in contact with it, but** all injurious effects **may be removed by** applying **oil to** the spot.

As a collector walks much, he must have something on his **feet that is easy and** at the same time serviceable. I **have found that** in stony countries like **New** England the best things are canvas **shoes** that **lace up in front,** tightly about the ankles and over the instep, to prevent slipping up **and down, which is the worst** possible thing that could **happen** while on **a long tramp ;** the soles should **be** broad, so that the toes may have room enough without crowding. With such shoes I have found that I could walk farther **than with anything else, and be less wearied in** the end. **If the** feet **are wet from** walking in **water,** with canvas shoes on them they will soon dry, **as the** water will all run out upon walking a short time **on** dry ground. Anything that is water-proof will be **much** too heavy to travel in, besides being injurious to the feet.

In sandy localities, or on marshes, or in winter when the **snow** covers the ground, Indian moccasons are the easiest and **best** things that can possibly be worn ; but in stony places they are not of sufficient thickness to protect the feet from receiving injury from the hard surface, otherwise they are exceedingly easy. They are not water-proof,

so that unless the snow is frozen in winter they are of no
use.   These moccasons can be procured almost anywhere
in Maine and New Hampshire, and sometimes in Boston.
They are manufactured mostly in Canada.   For clothes,
perhaps the best that can be worn in summer is a suit
of fine canvas of some dark color, to correspond with the
foliage; in winter, white, to correspond with the snow;
in both cases the wearer is less conspicuous, and can approach his game much more readily.   This cloth will not
wear out or tear easily, and is every way fitted for travelling in the woods.

I would next call attention to making stands on which
to put birds after they have been mounted, as one of the
necessities of the cabinet.   Simple stands in the form of
the letter T (Plate VIII. Fig. 3, f) are generally wanted.
Any carpenter can make them.   Different sizes will be
needed, from one with the standard two inches high with a
cross-piece one inch long, to a foot standard with a six-inch
cross-piece, with bottoms to match.   If made of pine, these
stands may be painted white, of a very pure unchanging
color, in the following manner.   Buy white zinc at thirty
cents per pound, and nice frozen glue at from twenty-five
to thirty cents per pound; dissolve the glue thoroughly in
hot water, then strain; to a pint and a half of water use a
quarter of a pound of glue, to this add one pound of zinc,
stir well, with the vessel that contains it in boiling water,
then, with a brush, apply to the stands; put on two coats.
If the paint has a yellowish cast, put in a few drops of
bluing; it will change it at once.   Thus you will find
that you have a nice white coat of paint that will remain
unchanged longer than oil colors.   Any other color can be
used, if preferred, in the same manner.

Fancy stands are made in the following way.   For mossy
stands, select a wooden bottom of suitable shape and size, —
those with the edges bevelled are generally used, — and

with the pliers force a piece of wire into it in the centre, then bend the wire in imitation of a branch or small tree, then wind it with hemp to give it the required shape; additional wires may be fastened on to represent the smaller twigs.

The whole is now to be covered with a coating of glue, and sprinkled with pulverized moss, or small pieces of moss are placed upon it smoothly. If the work is performed neatly, a perfect imitation of a little tree will be the result, upon which the bird is placed. If artificial leaves are to be used, they may be placed upon the twigs with glue.

If, instead of a wire, a twig bent in the required form can be procured, and fastened to the bottom with wire, it may be covered with moss without winding with hemp. The fancy stands seen with dealers in birds are generally made of a substance called *papier-maché*, that is, manufactured of paper pulp and glue as follows: Tear paper in small pieces and place it in water, let it stand overnight. Then, as it will be entirely soaked, reduce it to a perfect pulp, either by forcing it through a sieve or by stirring it. When reduced to a pulp, drain the water away. Dissolve a quarter of a pound of glue in a pint of water; mix with this a pint of pulp, heat it, and stir it well; then it is ready for use. Prepare a stand as described. Mould the pulp upon it in any shape to suit the fancy. It should have the consistency of putty, in order to work well. If it is too thin, put in more of the pulp; if too dry, more water.

With this substance you can imitate almost anything in the shape of miniature trees, with hollows, knots, crooked limbs, etc. By drawing over the whole, when finished, a comb, the bark of a tree can be imitated exactly. When perfectly dry, the limbs of the tree can be painted brown in the manner described. The bottom of the stand is painted green, and sprinkled with a substance resembling green sand, called "smolt," which may be procured at the painter's; over this is sometimes sprinkled thin glass, broken fine, which is called "frosting," and is also used by painters.

Rocks can be imitated well with *papier-maché*. If studded with small pieces of quartz the effect is heightened; they then may be painted in imitation of granite, sandstone, etc. The sandstone is easily imitated by sprinkling on sand before the pulp is dry. There are many other things that may be imitated with this wonderful substance, but, having given the preceding hints, I leave the reader to experiment at his leisure upon them.

For the scientific cabinet I would advise the use of the plain white stands as being much neater. The others are only fit for ornament.

One other thing is necessary. Take a thin board, and at intervals of two inches tack transversely strips of wood (Plate IV. Fig. 1); then cut a strip of paper as wide as the board, and with glue make it adhere at the top of the strips and at the middle of the intervening space, so as to form a corrugated appearance (Fig. 2). These are used in drying skins of birds. Each board should have about twelve such spaces, varying in width from two to four inches, the boards varying in width from four inches to one foot. These boards, with careful use, will last a long time.

Section III. *Measuring, Skinning, and Preserving Birds.* — For measuring, a pair of dividers, or compasses, a steel rule, divided into hundredths of an inch, and a longer rule, divided into inches and half-inches, will be wanted. To measure the bird, proceed as follows: Place the bird upon its back upon the longer rule, with the end of the tail at the end of the rule; the neck is stretched at full length, without straining; the bill must be pointed with

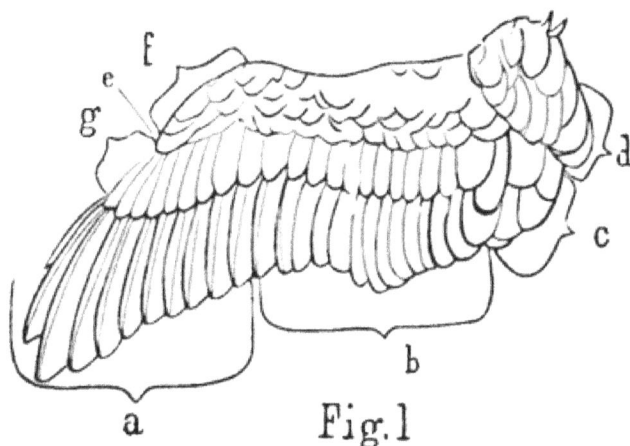

Fig. 1

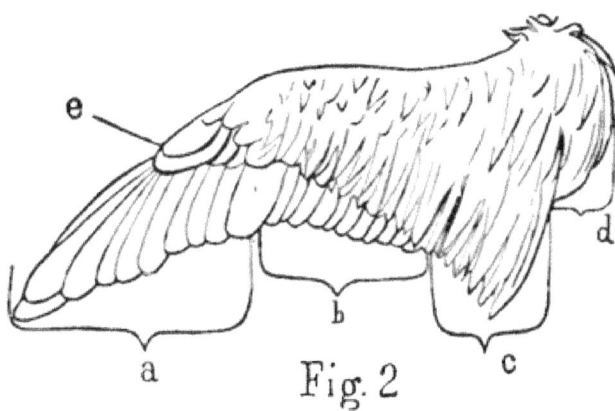

Fig. 2

Plate II.

the rule. Record the number of inches upon a strip of paper; if there is a fractional part of an inch, measure it with the dividers, and find how many hundredths it contains upon the smaller rule, and record it. This is "the length of the bird."

Stretch the wings out to the full length, with the bird still upon its back; measure these from tip to tip as "the stretch of wing." Measure the wing from the tip to the carpel joint, or bend, with the dividers (Plate X. d), for "the length of the wing." The tail is to be measured — also with the dividers — from the tip to the root for "the length of the tail." Measure the tarsus (Plate VII. g) as "the length of the tarsus." Measure the bill, from the tip of the upper mandible to the base (if the base is not well defined, as in the Ducks, measure to the feathers); this is "the length of bill along the culmen" (Plate III. h). Measure from the tip of the upper mandible to the gape (c) for "the length from gape"; also from the tip of the lower mandible to the angle of the gonys (s) for "the length of gonys" (c). In the Hawks, measure to the cere.

The color of the eyes, feet, and bill is now observed and recorded, also the date of collection and the locality in which the bird was collected. If the bird is in worn plumage, the fact should be recorded, as this will affect the measurements; also if it is moulting or in perfect plumage. As the records now made are only temporary, signs may be used to save time, such as X—— would denote an adult bird in perfect plumage, Y|—— would denote a young bird in worn plumage, YY|——| would denote a young bird not a year old and moulting, — this stage in the life of the bird is called the "young-of-the-year." By using some such signs as these much time will be saved. When the collector becomes expert at measuring, he will find that all small birds can be measured and recorded in

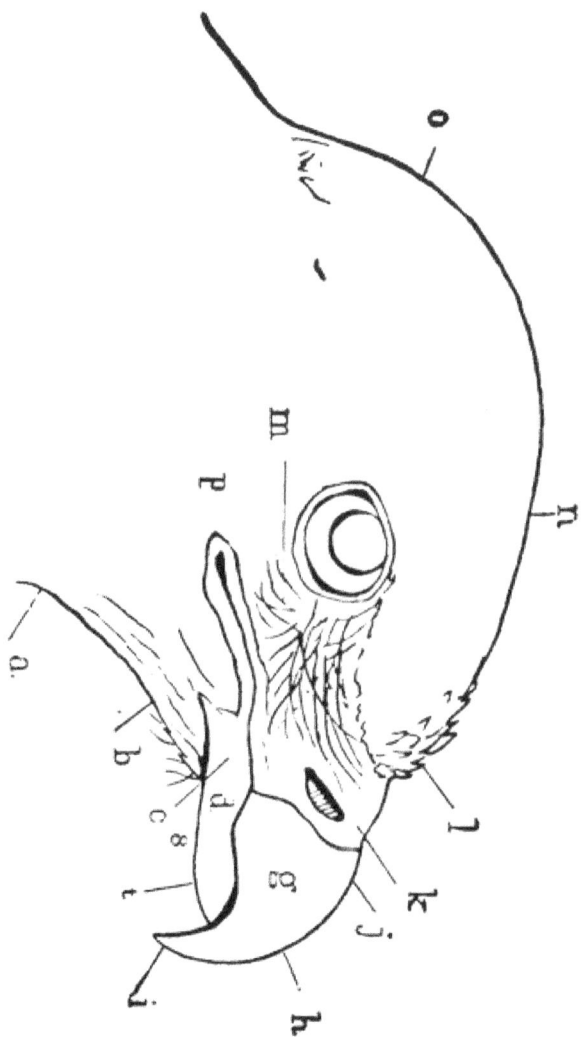

Plate III

about three minutes, and the larger ones in a little longer time.

Skinning. — First, have plenty of plaster near at hand. Remove the cotton from the mouth and vent, and place a fresh plug in the mouth alone. The method now about to be described is one that will apply to all birds, excepting those to be hereafter named.

Place the bird upon its back; with the forefinger and thumb part the feathers on the abdomen, and a bare longitudinal space will be discovered, extending from the breast to the vent. With the scalpel divide the skin in the centre of this bare space, commencing at the lower part of the breast-bone, or sternum (Plate X. o), and ending at the vent. Now peel the skin off to the right and left, and sprinkle plaster upon the exposed abdomen. Force the leg on the right side up *under* the skin, at the same time drawing the skin down until the joint (p) appears; cut through this joint and draw the leg out as far as the tarsus or first joint (k); with the point of the knife sever the tendons on the lower part of the leg, then by a single scraping motion upwards they may all be removed, completely baring the bone; treat the other leg in a like manner, leaving both turned out as they were skinned. Place the finger under the rump near the tail, then with the scalpel cut through the backbone just in front of the coccygus (u) entirely through the flesh to the skin, — the finger beneath is a guide to prevent cutting the skin. This may be done very quickly after long practice, and there is no danger of severing the skin if proper care be used. Put on a fresh supply of plaster. Now grasp the end of the backbone firmly between the thumb and forefinger, and with the other hand pull the skin down on all sides towards the head, until the joint of the wing, where the last bone, or humerus (r), is joined to the body, appears; sever the bones at this joint, and draw the skin down over the neck and head. When the ears appear, with the thumb-nail remove the skin that adheres closely to the skull without breaking it, pull down to the eyes, then cut the skin off close to the eyelids, taking care not to cut or injure them; but be sure and cut close enough to remove the nictating membrane, as it will otherwise cause trouble. Skin well down to the base of the bill. Remove the eye with the point of the knife by thrusting it down at the side between the eye and the socket, then with a motion upward it can be removed without breaking; cut off enough of the back part of the skull to remove the brains easily. Proceed to skin the wings; draw them out until the forearm (Plate X. d) appears, to which the secondaries are attached; with the thumb-nail detach them by pressing downward forcibly. Remove the muscles and tendons — as explained on the leg — to the joint, where the forearm joins the humerus (B), then divide, removing the humerus entirely.

Now open the drawer containing the arsenic, and with a small flat piece of wood cover the skin completely with it; be sure that the cavities from which the brains and eyes were removed are filled. Take up the skin and shake it gently. The arsenic that remains adhering to it is sufficient to preserve it, provided the skin is damp enough; if not, it may be moistened slightly. Now fill the eye-holes * with cotton, tie the wing-bones with thread, as near together as the back of the bird was broad, then turn the skin back into its former position. Smooth the feathers of the head and wings with the fingers. With a few strokes of the feather duster, holding the skin up by the bill, remove the plaster and arsenic that may be adhering to the feathers.

If there is blood upon the feathers, it may be removed — if there is not much of it, and if it is dry — with the

* By which I mean the holes occupied by the eyes in the skull.

stiff brush by continuous brushing, assisted by scraping with the thumb-nail. A living bird cleans blood from its plumage by drawing each feather separately through its beak, thereby *scraping* off the blood ; the thumb-nail performs the part of the bill. If much bloody, with a soft sponge and water wash away *all* traces of blood ; then throw plaster upon the wet spot, and remove it before it has time to harden or " set." By repeating this operation, at the same time lifting the feathers so as to allow the plaster to dry every part, and by using the soft brush, the feathers will soon dry. In this way any stains may be removed.

If the plumage is greasy, wash it with warm water and strong soap long enough to remove *every particle* of fatty matter that adheres to the feathers ; then rinse *thoroughly* in *warm* water, afterwards in cold. Be sure and remove all traces of the soap before putting on the plaster to dry, as the soap will be changed by the plaster into a gummy substance, which will be *very* difficult to remove.

After smoothing the feathers carefully, place the skin upon its back. With the tweezers take up a small roll of hemp or cotton, as large round and as long as the neck of the body that was taken out, and place it in the neck of the skin, taking care that the throat is well filled out ; then, by grasping the neck on each side with the thumb and finger, the hemp or cotton may be held in place, and the tweezers withdrawn. After placing the wings in the same position as the bird would have them when at rest, with the bones of the forearm pushed well into the skin, — so that they may lie down each side, and not cross each other, — with a needle and thread sew through the skin and the first quill of the primaries by pushing the needle through the skin on the *inside* and through the quill opposite, but be sure that the wing is in the proper place. (If it is too far forward, the feathers of the sides of the breast, that ought to lie smoothly over the bend of the wing, will be forced up and backward. If the wing is placed too far *back*, there will be a bare spot upon the side of the neck, — caused by the wing-coverts, which help, in connection with the feathers of the back, to hide the spot, being drawn either down or back too far. If the wing is placed too low, the same spot is seen, only it is elongated and extends along the back between the secondaries and feathers of the back ; if too high, the feathers of the back will appear pushed up, and will not lie smooth for obvious reasons. When the wing is in the right position, the feathers of the wing-coverts and back will blend nicely and smoothly, and the feathers of the sides of the breast will lie smoothly over the bend of the wing ; the ends of the closed quills will lie flat upon the tail, or nearly so.) Now draw the thread through so that but an inch is visible inside the skin, then push the needle through the skin from the outside just *below* the quill that it came out through, draw the thread through, and tie to the projecting end, thereby fastening the wing firmly to the side ; proceed in this way with the other wing.

Roll up loosely an oblong body of cotton or hemp of the same size as the body taken out, place it in the skin neatly, then draw the edges of the skin together where the incision was made, and sew them once in the centre ; tie the ends of the thread together. Take care to put the needle through the *edge* of the skin so as not to disturb the feathers. Smooth the feathers on the abdomen. Cross the feet upon the tail (Plate IV. Fig. 3), — which is spread slightly, — then place the skin upon its back in the rounded places of the drying-board, spoken of on page 18 (Fig. 1, d), taking care that the feathers of the back are perfectly smooth. This rounded bed gives the back a natural rounded appearance, which cannot be made easily in any other way. Place the head with the

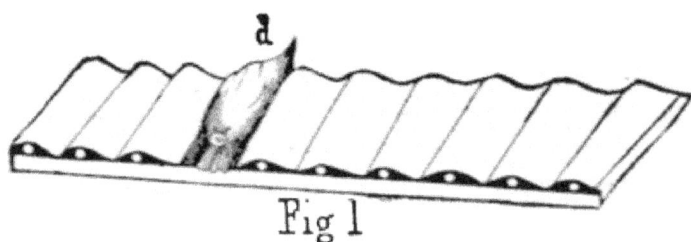

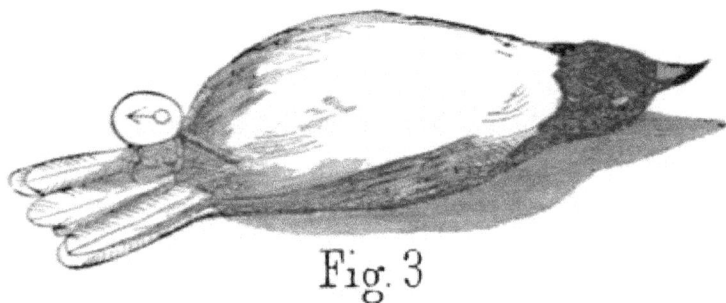

Fig 1

Fig. 2

Fig. 3

Plate IV.

bill horizontal with the back or bottom of the rounded
space, with the culmen (Plate IV., Fig. 1, d) nearly touch-
ing the paper. The skin must remain in this position
without being disturbed until perfectly dry, which in very
warm weather, with small birds, will be in about twenty-
four hours.

If this corrugated board cannot be procured, the skin
may be placed on its back upon a flat surface, with a
little cotton on each side of it to prevent its getting dis-
placed. This is what is technically termed "a skin"
(Fig. 3), and this method of making them is the best I
have ever seen practised, and one that I have used for
years as being the most expedient. The skins so made
are less liable to injury, being stronger than some others,
and are also very easily mounted. I have made in a
single day, in the manner described, fifty skins, and with
practice almost any one will be able to do the same; ten
minutes being ample time for each, including the measuring.

Before the skin is placed upon the board, it should be
labelled (Fig. 3, ♂) with a number corresponding to the
one placed upon the slip of paper containing the measurements, etc., marked also for the sex of the bird, which
is done by using for the male the sign of the planet
Mars, thus ♂; for the female the sign of the planet
Venus is used, thus ♀. These signs are used by naturalists throughout the scientific world, and it is best to become accustomed to them.

Determining the Sex.—The sex of the bird is determined,
not by the plumage, which will sometimes set the student
at fault by its changes, and *should never be trusted in determining the sex*, but by dissection, as follows: Take the
body of the bird after it has been removed, and cut with
the scalpel through the ribs (Plate X. A) on the sides
of the abdomen, thereby exposing the intestines; raise

these gently with the point of the knife, and beneath them
will be seen the sexual organs, which are fully illustrated
in the following diagrams.

PLATE V., Fig. 2, is an adult male ($\male$) in the breeding
season. 1 shows the position of the lungs, 2 the peculiar yellowish glands, — in some birds bright yellow, in the
present case — that of a song sparrow (*Melospiza melodia*,
Baird) — they are yellowish white, which, being present
in both sexes, if not examined closely, may be easily mistaken, in the young female, for the testicles of the male.
3, 3, are the testicles, much enlarged in this, the breeding
season. The sex of a bird in this stage is easily determined.

PLATE VI., Fig. 1, is a young male ($\male$) in the young-
of-the-year plumage. The figures refer to the same parts
as explained in the preceding. It will be perceived that
the testicles (3) are much smaller. At different seasons,
the testicles vary in size between this and the preceding.
In some birds they are elongated, and black in color,
as in the Herons; but they always occupy the same positions so nearly as to be readily distinguished. The peculiar white glands (2) are in this instance very prominent,
but they are readily known in all birds by their being flat,
while the testicles are always spherical.

PLATE V., Fig. 1. This is an adult female ($\female$) in the
breeding season. 1, 1, are the same peculiar glands observed
in the males; 2 is the ovary, a mass of spheres at this
season quite yellow and prominent; 3 is the oviduct, or
egg-passage, much enlarged in the present case, as it always
is during the breeding season, when it assumes a thick,
swollen appearance, while at other times it is translucent,
much smaller, and resembles a narrow, whitish line, not
readily perceived.

PLATE VI., Fig. 2, is a young-of-the-year female in autumn. 1, 1, the same white glands that at this stage of the

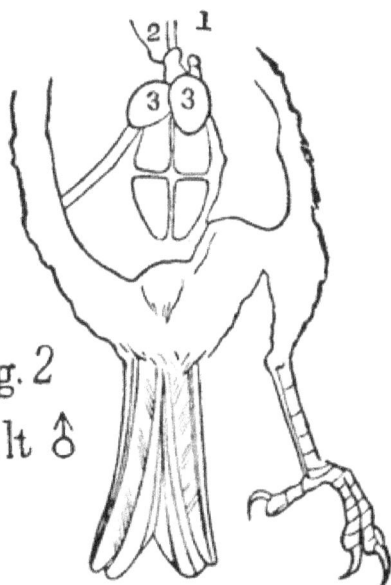

Fig. 2
Adult ♂

Fig. 1
Adult ♀

Plate **V.**

Fig. 2
Young ♀

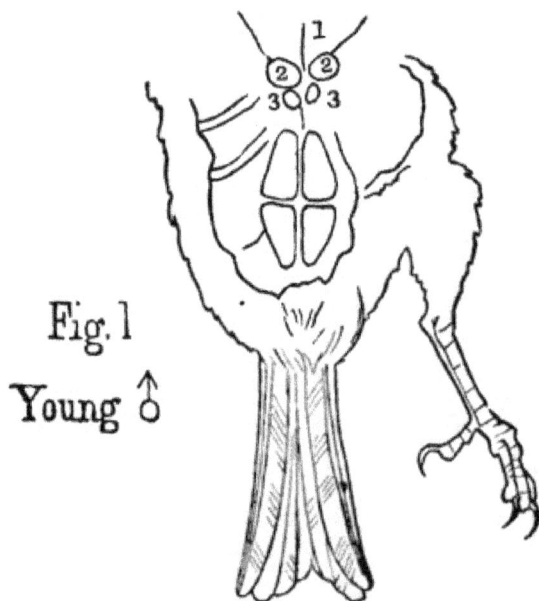

Fig. 1
Young ♂

Plate VI

bird's life might at first sight be mistaken for the testicles
of the male, but, upon looking closely the ovary (2) can
be perceived, very small; upon applying a magnifying-glass
it appears granular.

With these remarks and diagrams, I think any one with
ordinary ability will, with a little practice, be able to de-
termine this very important character in the scientific
study of birds.

Contents of Stomach, etc. — The contents of the stomach
must next be examined, which is done by opening the giz-
zard and crop. A little practice will enable the collector
to state correctly what it contains, although the glass is
sometimes necessary, as in the case of small birds. This
is then recorded upon the slip of paper, which is put on
file, to be copied into a book, in the following manner, —
leaving a page, or, if the book is not wide enough, two
pages, for each species, — first placing the *scientific* name at
the head, as seen on the following page.

A book prepared in this manner, carefully indexed and
paged, will, when it is filled with the measurements of
birds, be of immense value for comparative measurements,
besides giving the collector a complete history of each of
his specimens.

Exceptions to the usual Method of Skinning. — All birds
are to be prepared in the preceding manner, with the following exceptions.

All Woodpeckers with a large head and small neck — of
which the Pileated Woodpecker (*Hylotomus pileatus*, Baird)
is an example — should be skinned in the same manner as
far as the neck, which should be severed, as it is impossible to turn the skin over the head ; cut through the skin
on the back of the head, making a longitudinal insertion
of an inch or more, and draw the head through this. It
should be carefully sewn up after the skin is turned back.
Such specimens, when laid out to dry, should have the

Sialia sialis.

No.	Date.	Locality.	Age.	State of Plumage.	Sex.	Length.	Stretch of Wing.	Wing.	Tail.	Bill.	Tarsus.	Color of Eyes.	Color of Bill.	Color of Feet.	Contents of Stomach.	Remarks.
1022	1868. July 22	Newton, Mass.	Young-of-year	Perfect	♂	7.00	12.45	3.20	2.56	.50	.82	Brown	Brown Black	Black	Beetles and Flies	Spotted on breast.
1020	"	"	Adult	Worn	♀	7.00	12.00	3.96	2.40	.52	.62	"	"	"	Beetles	Shot in a field.
1934	Dec. 30	Jacksonville, Fl.	"	Perfect	♂	7.00	12.30	4.00	2.75	.49	.80	"	"	"	Berries	Shot on the Pine Barrens.
1969	1869. Jan. 3	"	"	"	♀	6.60	11.75	3.75	2.65	.50	.75	"	"	"	Seeds	Shot on the Pine Barrens.

head so placed that the bill is turned at right angles with the body, with the head resting on *one side*, and not on the *back* as before.

All Ducks with large heads should be skinned in the same way, with the exception that the insertion must be made *under* the head, on the throat. Ducks, Herons, Geese, large Sandpipers, and all other long-necked birds, should, when placed to dry, rest upon the *breast*, with the head and neck placed upon the back, and the head turned on one side. Herons with very long necks should have them bent once. The bill must be placed parallel with the neck and pointing forward.

While travelling it is not always convenient to fill out the bodies of large birds; it is better to pack them flat, with but little cotton in them, — just enough, however, to keep the opposite parts of the skin from coming in contact. The neck should *always* be filled. When it is impossible to procure stuffing for small birds, they may be packed flat also.

Birds that have been preserved with carbolic acid, even after they have been lying for years, and have become perfectly dry, may be skinned in the following manner: Place wet cotton or hemp under the wings, in the throat, and around the legs, and finally envelop the whole body in a thick coating. Place it in a close box, and let it remain a day or two until it is softened, then take it out and remove the skin as before directed; but more care will have to be exercised than in skinning fresh birds. Alcoholic specimens may be skinned; but the wet plumage had better be dried in the air without plaster. Mr. A. L. Babcock has a number of mounted birds in his collection at Sherborne, Massachusetts, that were preserved in alcohol, and sent from South America.

The beginner will find that some birds, such as the Cuckoos, Pigeons, and Doves, are very difficult to skin over

2*   c

the rump without loosening the feathers; but this difficulty will be overcome by using particular care while skinning the spot that is tender. Some birds also have tender skin on the breast, and in such cases it almost always adheres so closely to the flesh that it is necessary to cut it away; this operation is somewhat tedious, but it is better than to risk tearing the skin by pulling it. An example of this is sometimes seen in the Wood Duck.

The best time to skin a bird is as soon as it is shot, when the muscles are relaxed, as the plumage is then in the best condition. In a short time the muscles become rigid, when it is extremely difficult to remove the skin; but the muscles soon relax again, and then you must skin *at once*, as this is the first stage — or the state immediately preceding the first stage — of decomposition. In very warm weather this rigidness of the muscles seldom occurs, then the bird rapidly decomposes. In warm weather, birds should be kept on ice until wanted; for if a bird remains only an hour in a warm room, or in the sun, it will sometimes spoil, especially if the blow-flies are allowed access to it. There is a species of blow-fly that is viviparous; I have seen such a fly alight upon a fresh bird, and, after introducing her ovipositor into the mouth of the specimen, exude an immense number of living, though minute, maggots. These maggots spread over the skin in all directions, moistening it with their slimy bodies, and soon render the specimen unfit for use by loosening the feathers.

It is difficult to remove the eggs of the common blow-fly when they are once placed upon the feathers. It is much better to prevent the flies from attacking specimens — which, if they are exposed during warm weather, they will do very quickly — by covering them, or placing them immediately upon ice.

SECTION IV. *Mounting Specimens.* — Almost any one

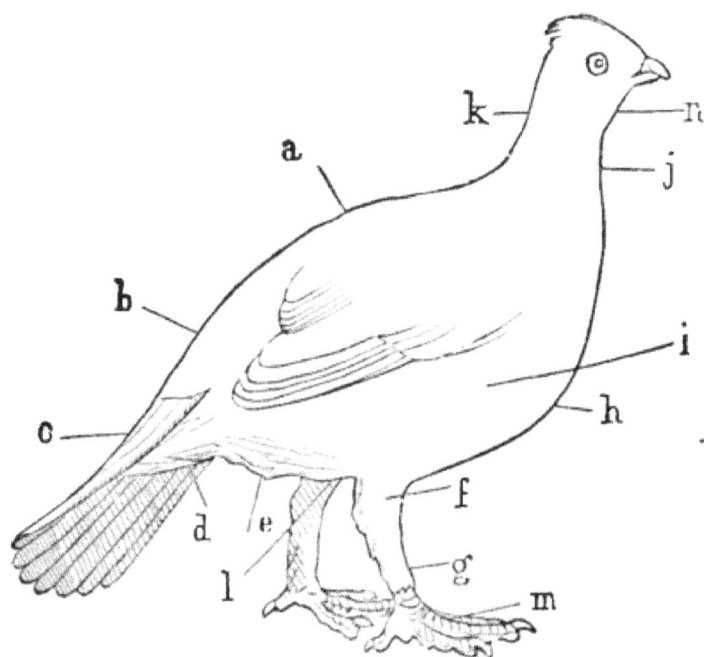

Plate VII.

can mount a bird, after receiving proper instruction ; but to make it look lifelike and natural requires constant and unceasing study of birds in their native haunts. The true art, then, can only be acquired by the earnest student of nature. The mere taxidermist, who constantly sits at his bench and works on birds without studying from nature, may acquire a certain degree of *skill*, but the attitudes of many of his stuffed birds will appear awkward and grotesque to the *naturalist*.

Therefore I say to those who would learn to mount birds in natural attitudes, *study nature*. Have all attitudes that every bird assumes engraved upon the brain, to be reproduced in the stuffed specimens ; from the one assumed by the delicate Warbler, that hops lightly from limb to limb, or swings gracefully from the topmost bough of some tall oak, to that of the mighty Eagle in his eager, downward swoop upon his trembling prey. Watch the screaming Gull in his almost innumerable positions upon the wing, the nimble Sandpiper running along the shore, and the gracefully floating Duck upon the water. After watching these in their various natural attitudes, work ; but do not cease to study for improvement, for the work of man is yet far from being *perfect*.

In mounting birds, skin as instructed in the preceding section, but do not tie the wing-bones together. Having cleaned and dusted the feathers, proceed to fill the neck to the natural size, without stretching, with "shorts," or the bran from wheat flour, or with hemp cut fine. Roll up some fine grass moderately hard in the shape of an oblong body (Plate VIII. Figs. 1, 2), then wind it smoothly with thread. This body should be of the same proportionate size as the one taken out, although not exactly of the same shape, for reasons that will be seen when the bird is mounted, but which cannot be easily explained. Have the body perfectly smooth, and the curves regular on every part.

Place the body inside the skin. Now cut wires of the right size (that is, large enough to support the bird when mounted; which can be learned by experience, although I would advise putting in as large wires as can be used without splitting the skin of the tarsi) and proper length for the wiring of the following parts : to go through the legs, for the neck, and for the tail. Straighten the wires by rolling them on the bench with a file, then sharpen them by holding the end obliquely against the edge of the bench and filing from you, at the same time twisting the wire; force the wire cut for the leg up through the sole of the foot, through the tarsus, along the leg-bone into the centre of the side of the grass body (Plate VIII. Fig. 1, a), through this so that the end will protrude for an inch; bend the end down in the form of an L (Fig. 1, b), and again force it into the body (Fig. 1, c), thereby clinching it so that it can have no motion whatever. The wire should protrude out of the sole at least two inches (Fig. 1, d).

Proceed in the same manner with the other leg; if this seems difficult at first, practice will soon overcome the difficulty. Be sure and clinch the wires *firmly*, as they will otherwise cause trouble. Force the wire cut for the head down through the skull near the base of the bill (Fig. 3, g) through the neck, — but it must not come out through the skin anywhere, — through the body (Fig. 1, e), out the other side, where it is clinched as before (Fig. 1, c). Force the wire cut for the tail through the bone left in the tail, and under the tail, into the body (Fig. 1, f); clinch as usual (Fig. 3, g). Bend the wire — which should protrude about two inches beyond the end of the tail — into the form of a T (Fig. 1, h); the cross-piece is placed about half the length of the tail; on this the tail rests. Pin up the incision by drawing the edges of the skin together and forcing pins through them into the body; then smooth the feathers over the place.

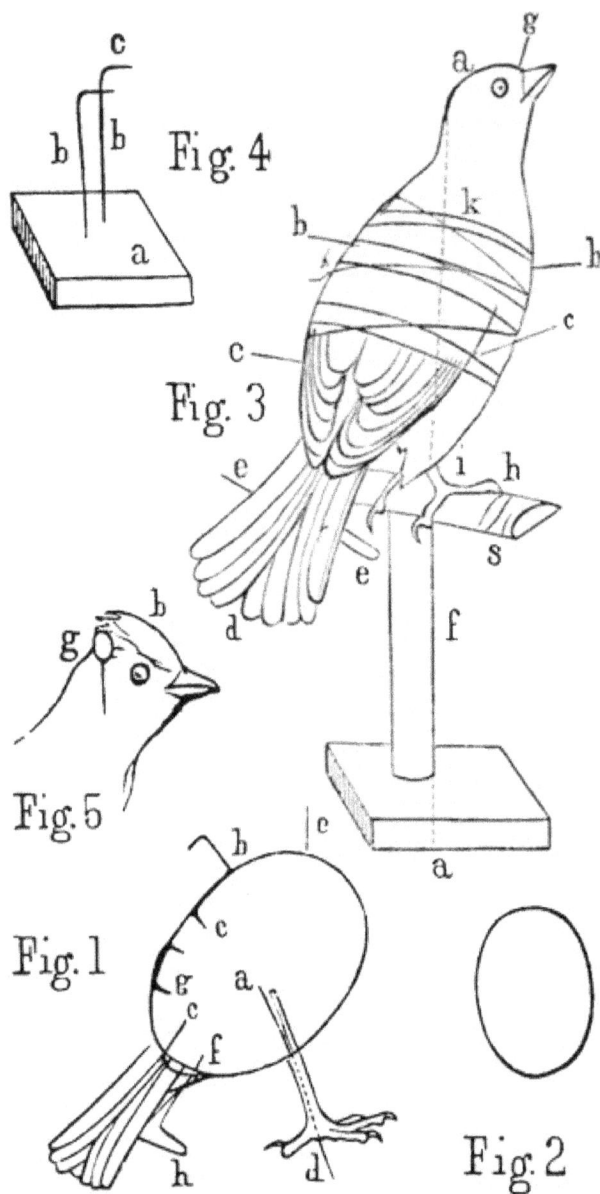

Fig. 4

Fig. 3

Fig. 5

Fig. 1

Fig. 2

Plate VIII.

Fasten the bird upon the stand by passing the wires of the feet through the holes in the cross-piece (Plate VIII. Fig. 3, h), then twist the ends of the wire around the ends of the cross-piece (Fig. 3, s) to fasten it firmly. Place the bird in position with the tarsi inclining backward (Fig. 3, i), so that a line dropped from the back of the head, passing through the body, would pass down the centre of the stand (Fig. 3, a, a). This is a natural rule, and one that applies to all perching birds. Next arrange the wings in position by applying the same rules that were given when making a skin; in this case, however, the following additional rule may be of use. The end of the bone of the forearm should reach just half the length and width of the body where it meets the lower end of the thigh (see Plate X. B). Having arranged the wing, pin it near the bend to the body (Plate VIII. Fig. 3, b), also through the first primary quill (Fig. 3, c). The wings should in some cases be placed at a little distance from the body, as is natural with the Thrushes, and some other species. This may be done by lengthening the second wire (Fig. 3, c). Put the head in the proper position, cut off the protruding wire (Fig. 3, g). Plait the tailfeathers by placing the inner web over the outer (Fig. 3, d); then place a piece of fine copper wire across the tail, and fasten it to the ends of the cross-piece (Fig. 3, e). Fix the artificial eyes in their proper position with glue or putty; then wind the bird with the fine cotton on the breast and shoulders and over the secondaries (Fig. 3, k); this is to keep the feathers smooth while it is drying.

To mount a bird with the wings extended, proceed as before explained, but raise the wings, and use longer wires to pin them in position. Then, to hold the quills and secondaries in place, bend a wire over the whole width of the wing, passing on each side of them. In skinning and mounting Ducks, open under the wing. This is accomplished by making an incision on the side, from the place where the humerus joins the sternum (Plate X. D) to just beyond the lower joint of the thigh (p), after which skin as before.

If any feathers become twisted or bent, they may be instantly straightened by holding them in steam. If the feathers are to be smoothed, raise them with the fingers or tweezers, and let them fall back in place; they will generally come down smoothly. If the feathers come out, put a drop of glue upon the end of each, and place it in the proper position; it will stay, when dry. In this way large bare places may be covered.

Mounting dried Skins. — To mount dried skins, remove the stuffing with which they are filled, and supply its place with dampened cotton, also wrap the legs well with it; place the skin in a box, where it must remain until it becomes pliable, but not too soft, as it is then liable to drop in pieces. It is to be mounted as described, excepting that the neck is filled with cut hemp instead of bran. Birds mounted from dried skins require more care in mounting, and more binding to bring the feathers into proper position, than fresh birds.

Never paint or varnish the feet or bill of a bird; the scales on the feet of birds are one of the most interesting characters in the study of ornithology, and they cannot readily be seen when covered with paint or varnish. In mounting birds larger than a Robin, the muscles of the leg-bone must be supplied by winding the tibia with hemp until the original size and shape is attained. Particular attention must be paid to the legs of the Waders and rapacious birds. The exceptions to this rule are all swimming birds; as the tibia is buried in the body, it does not need winding.

In mounting Humming-Birds with the wings extended, especially from dried skins, there is no need of wiring the leg. Place a single wire in the back part of the body, with

the point firmly clinched, and the end protruding back
from the abdomen for three or four inches. This wire will
sustain the bird. Always wire the feet in the usual way
if the bird is to be mounted in the attitude of rest. In
mounting other small birds, this method of wiring will answer when the bird is represented as flying. Large birds,
when mounted in ,the attitude of flying, should be wired
in the usual manner, with the wires that extend beyond
the soles of the feet cut short ; then a wire is forced down
through the back and clinched under the body, with the
end pointing upward ; cut off this end so that it will protrude but a half-inch beyond the skin, then bend it under
the feathers into the form of a hook or ring ; to this fasten
a thread, and suspend the bird. To make the bird incline
downward, place the wire well back ; upward, farther forward, or nearer the head. Fine copper wire will answer
to suspend large birds in this manner. In mounting a
bird in this position, with the wings fully extended, care
should be taken that they are properly arched.

While arranging the wings, it is not convenient to keep
the bird suspended, as it will not be sufficiently firm.
Sharpen two stout wires and fasten them at both ends
(Plate VIII. Fig. 4, b, b), perpendicularly in a block of
wood (a), parallel to each other, and about two inches
apart. These wires should be at least four inches long.
Bend about an inch and a half of the ends down, parallel
with the block (c). Force these ends into the abdomen of
the bird that is being mounted, and it will be firmly held
in place while the different parts are being arranged, after
which it can be suspended as described.

Sometimes it is necessary to mount the skins of rare
birds when they are badly decayed. To mount skins in
this condition requires skill and patience, as well as a knowledge of their different parts. The manner in which I have
mounted them is this : First, moisten the skin as de

scribed, then make a body, as before, and place a wire of the proper size and length in the usual place for the neck, and wind it with hemp to the natural size ; place the head, wings, feet, and tail in the proper position ; then, after spreading glue upon the body, place each feather or piece of skin carefully in its proper place, commencing at the tail and working towards the head ; when this is finished, bind the bird as usual.

Birds that have been mounted require at least ten days for the skin to dry before the thread is removed, which is done by cutting down the back with scissors, after which cut off all protruding wires, and unplait the tail-feathers and smooth them. To elevate the crest of a bird, or any other naturally elevated feathers, — such as the elongated feathers on the necks of some species of Grouse, etc., — roll a piece of cotton into a ball, and force a pin or piece of sharpened wire through it (Plate VIII. Fig. 5, g). Place this wire or pin in the bird in such a manner that the feathers to be elevated may rest on the cotton in a natural position (Fig. 5, b). After the skin becomes hardened the cotton may be removed, and the feathers will retain the desired position.

CHAPTER II.

SECTION I. *Collecting.* — Because mammals are not quite as interesting at first sight as birds, the study of this class of animals has been somewhat neglected; and but comparatively few naturalists are even aware of the existence of some of the smaller mammalia that live about them. They are, however, worthy of special attention, and, if studied, will soon be found particularly interesting.

In collecting mammals, excepting some of the larger species, the gun is of but little use; they must be taken almost entirely with traps. Shrews and Moles may be frequently found where they have been dropped by cats, who catch them, but do not eat them. In this way a great many valuable specimens may be obtained.

Another way to procure Mice and Shrews is to turn over old logs and stumps, under which these little animals frequently hide, and while dazzled by the light's coming in suddenly upon them, they may be readily seized in the hand. Mice may also be trapped.

Squirrels may be shot or caught in traps; Foxes, trapped, shot, or dug out of their holes. Woodchucks are easily trapped or dug out. Skunks are very disagreeable animals to handle, but when one once becomes accustomed to capturing them he can do it in perfect safety. The best way is to catch them in a "box-trap" baited with the head of a chicken; when caught, immerse trap and skunk in water until the animal is dead. Treated in this manner, they will not emit any of their disagreeable scent. By breaking the backbone with a stout stick, when the animal is caught in a steel tråp, the disagreeable emission will be prevented. All animals should be killed either by breaking the backbone or by compressing the ribs, to stop the breath ; *never* by a blow on the head, as this is liable to injure the skull, which must be preserved entire for scientific investigation.

The following animals may be decoyed into traps by means of peculiar scents : Foxes, Fishers, Martens, Minks, Weasels, Wildcats of all species, Otters, Beavers, Bears, Muskrats, and Raccoons. These scents are made of different substances. The musk of the Muskrat, contained in two glands situated just below the skin upon the back part of the abdomen, will decoy Muskrats and Minks, and perhaps Wildcats. This musk may be procured from the male in early spring. After the two glands spoken of are removed, they may be cut open, when the musk — which is a milky fluid — will appear, and may be squeezed out, mixed with alcohol, and kept for use. This musk is used in the following manner : Cut a stick of pine about six inches long, make a small cavity in one end ; into this drop a little of the musk, fasten the stick in such a position that the animal to be decoyed must place his foot upon the trap in order to reach it.

Foxes, I have been informed by old trappers, are readily decoyed by using the fetid scent of the Skunk in the same manner. This scent is a greenish fluid, and is contained in glands situated in the anal region ; it may be obtained in the same manner as the musk, although the operation is not pleasant. All of the above-named animals may be successfully decoyed by using an excessively fetid scent prepared during warm weather in the following manner: Take a good-sized eel or trout, and cut it in small pieces ; place it in a quart bottle, cover the top with gauze to keep the flies out, hang the bottle on the south side of a fence or building, and let it remain two or three weeks, when the whole mass will become decomposed; then on the top will be found a thin layer of a clear liquid having an ineffably disagreeable odor. This fluid should be poured off carefully into a small phial and closely corked; it is to be used in the same manner as the other scents.*

Bats may be shot, or taken during daylight beneath the shingles of buildings, or in hollow trees. One or two species, however, remain outside, suspended to a branch or leaf of a tree.

Plaster may be used to absorb the flow of blood from mammals, as well as from birds.

SECTION II. *Measuring.* — To measure a mammal preparatory to skinning: Place it upon its back, then with the dividers measure the distance from the tip of the nose to the front side of the eye, record this as "the distance from the tip of the nose to the eye," then from the tip of the nose to the ear; this is "the distance from the nose to the ear"; then from the tip of the nose to the occiput, or back of the head, for "the distance from the nose to the occiput." With the rule find the distance from "the nose to the root of the tail," also the distance from "the tip of the nose to the tip of the longest toe of the outstretched hind leg"; then with the dividers find the length of the vertebra of the tail from the root; this is "the length of the tail to the end of the vertebra." With the dividers, measure the hair on the end of the tail for "the length of the hair." Measure the length of the hind leg from the knee-joint to the tip of the longest claw of the longest toe for "the length of the hind leg." Measure the length of the front leg from the elbow-joint to the tip of the longest claw of the longest toe; this is "the length of the front leg." The width of the hand is found by measuring the width of the outspread forefoot or

---

* This receipt was kindly given to me by Mr. George Smith of Waltham, who has used it successfully, as I have personally witnessed.

hand. Now measure the length of the ear on the back side, from the skull to the tip, for " the length of the ear." Measure "the width of the muzzle" between the two nostrils. In animals larger than a gray squirrel, measure the "girth" with a tape-measure, or piece of string, just back of the forelegs.

These measurements will answer for all excepting the bats, in measuring which proceed as before ; but, instead of the forelegs, find " the length of one wing," " the length of the wing to the hook, or thumb," and " the stretch of wings " as in birds.

Seals also vary slightly ; instead of the word " leg " use " flipper," and find the width of the hind flipper as well as the width of the fore one ; also, in addition, " the distance between the fore-flippers." The sex of a mammal is easily determined without dissecting. These measurements are to be first recorded upon a strip of paper, as in the birds, and afterwards copied into a book, as seen on the next page.

Skinning. — To skin a mammal, place it upon its back ; make a longitudinal incision in the skin over the abdomen, extending from the root of the tail about one fourth of the length of the body. Peel down each side, as in skinning a bird, pushing forward the leg so as to expose the knee-joint; sever the leg from the body at this place, and clean the bone ; proceed in this manner with the other leg. In small animals, sever the tail as close to the body as possible, leaving the bone in ; but in large animals it can generally be removed by placing two pieces of wood on each side of the bone against the skin, holding them firmly in place with one hand, and after giving a strong pull with the other the tail will slip out easily. With some animals, such as the Beaver, Muskrat, Skunk, etc., this cannot be done ; then the skin of the tail has to be opened the whole length, and the bone removed. Proceed to draw the skin

Arctomys monax.

Locality.	Age.	Sex.	Date.	No.	Nose to Eye.	Ear.	Occiput.	Root of Tail.	Outstretched Hind Leg.	Tail to End of Vertebra.	End of Hair.	Hand. Hind Leg.	Length.	Width.	Height of Ear.	Muzzle.	Girth.	Skull. Length.	Width.	Remarks.
Ipswich	Adult	♂	Aug. 22 1868.	68	1.60	2.96	2.30	13.00	16.00	4.98	8.00	3.10	2.10	.78	.85	.20	—	—	—	Light colored.
"	"	♀	"	20	65	1.67	2.80	3.45	16.50	20 15	4.50	8.75	2.80	1.85	.92	.75	—	14.50	—	—	" "
"	"	♀	"	13	43	1.82	2.94	3.46	15.25	19.50	6.45	7.00	2.95	2.06	.70	.65	.16	9.75	—	—	Top of head black.

* This measurement is taken after the animal is skinned; the width of skull is measured on the widest part, the length on the longest part.

down towards the head, until the forelegs appear; sever these at the knee-joint, and clean the bone as before. Draw the skin over the head, cutting off the ears close to the skull. Use caution in cutting the skin from the eyelids and in severing the lips from the skull, so as not to injure their outward appearance. The skull is to be detached entirely. Cover the inside of the skin well with arsenic, and, if large, rub it in well with the hand; but be sure that every part is poisoned.

If there is any blood upon the hair, after the skin is turned into its former position, if it is dry, remove it with the stiff brush; if wet or *very* bloody, wash and dry with plaster, as explained in birds.

Wind the leg-bones with sufficient hemp or cotton to supply the place of the muscles; then fill out the head, neck, and the rest of the body to their natural size. Sew up the orifice through which the body was removed neatly over and over, drawing the edges of the skin together nicely.

Label the skin by sewing a bit of card-board upon one of the feet, or, if the animal is large, upon the ear, with the number of the specimen and the sex marked upon it. Clean the skull as much as possible with the scalpel; if it is a large animal, the brains may be removed through the orifice where the spinal cord enters the skull. If this opening is not large enough to remove them, they should be left in. Roll the skull in arsenic, then label it with a number corresponding to the one upon the skin, and lay it by for future cleaning. The arsenic prevents insects from attacking it.

Place the skin, if a small one, upon its side, with the legs bent neatly; if a large one, upon its breast, with the legs stretched out on each side, the forelegs pointing forward, the hind ones backward. This is what is technically called a " mammal's skin."

Very large animals, such as Deer or Bears, should not be filled out in this way, but placed flat. In skinning large animals, make an incision in the form of a double cross, by making a longitudinal cut between the hind legs, from the root of the tail to the breast, between the forelegs; then a transverse cut from the knee of the foreleg down the inside of the leg to the opposite knee. The same operation is repeated upon the hind legs. Then proceed as before, only, when the skin has been removed from the flanks, the animal must be suspended to facilitate the removal of the rest.

In skinning a mammal with horns, make a longitudinal incision from the back of the neck to the occiput, or back of the head; then make a transverse cut across the head, commencing about four inches beyond the right horn, and ending about four inches to the left of the left horn, the cut passing close to the base of the horns, thus forming a **T**. Remove the skin from the body as far as the neck, which is cut at its junction with the body. The skull, horns, and neck are drawn through the above-mentioned orifice.

In skinning large animals, it is well to take the diameter of the eye before it is removed, so that an artificial one may be inserted of the same size, if the animal is to be mounted, as the eyelids shrink very much while drying. All mammals should be skinned as soon as possible after they are killed, especially small ones, as in a few hours decomposition will commence; then the hair will come out.

While skinning the legs of ruminants, such as Deer, Sheep, etc., it will be found that the skin cannot be drawn over the knee-joint; then cut longitudinally through the skin below the knee, and after severing the bone at the hoof and knee, remove it through this incision. The incision should be about one fourth the length of the distance from the knee to the hoof.

Bats are to be skinned in the ordinary manner, remov-

3   D

ing the skin even to the tip of the phalanges of the wings; then tie the wing-bones together, as explained in birds. Place the bat upon a flat board to dry, and pin its wings in the proper position for flight. When dry, stitch it upon a piece of card-board.

While skinning mammals, it is sometimes necessary to use plaster to absorb the blood and other juices that are apt to flow; but if care is taken not to cut the inner skin over the abdomen it will not be needed. It is also sometimes necessary to plug the mouth and nostrils, especially if blood flows from them.

SECTION III. *Mounting Mammals.* — The art of mounting mammals in lifelike attitudes can only be acquired by experience. Hence the learner must practise the utmost degree of patience and perseverance. As in the first chapter I earnestly advised those who would be perfect to study nature, I would here repeat that advice. And if necessary while endeavoring to mount a bird, where the feathers cover the minor defects, it is essentially much more of a necessity to study nature carefully while striving to imitate the graceful attitudes and delicately formed limbs of the smaller species of mammalia, or the full rounded muscles and imposing attitudes of the larger ones; for in mammals the thin coat of hair will tend rather to expose than hide the most minute imperfections.

Perfectly stuffed specimens can only be obtained by careful measurements of all the parts, such as the size of the legs, body, etc.

In skinning mammals to mount, it is best not to remove the skull. Open it on the occipital bone, so as to remove the brains; clean well; cover with arsenic; then supply the muscles removed, by using hemp wound tightly on with thread. As the skin will shrink badly if it is stuffed loosely, carefully fill out the space occupied by the muscles of the legs in the same manner. Cut wires for the

feet, head, and tail, sharpen them on one end as directed
in mounting birds; now roll up grass until it is not quite
as large round as the body, and about one third as long.
Fill the fore part of the skin with bran or cut hemp as far
back as the shoulders, and place the ball of grass against
this filling, inside the skin. Now force the wires through
the soles of the feet and top of the head into this ball;
clinch them firmly. After filling the skin of the tail with
bran, force the wire through the grass ball to the very end;
then clinch the opposite end in the ball by cutting off the
part that protrudes and turning it in.

Fill the remaining parts of the skin with bran to the
natural size, and sew up the orifice carefully; place the
animal in the proper position by passing the protruding
wires of the feet through holes in a board, clinching them
firmly on the under side. The skin may now be moulded
into shape with the hands, the hair carefully smoothed,
the eyes set in the head with putty, the protruding wires
cut off, and the specimen set away to dry. There are
but few rules to be followed in placing animals in position, because they are almost infinite in variety. The most
imperative rule applies to the positions of the legs, which
are almost always the same; and it should be studied with
particular care, as a slight deviation from it will impair
the lifelike attitude of the specimen. The rule is: Never
place the bones of the first joint (Plate IX. No. 1) and
those of the second joint (2) of the hind legs in a *straight
line*, but always at an *angle*, more or less; while the two
bones of the forelegs (3, 4) should almost always be placed
in a straight line, — *always* when the animal is standing
upon them.

In imitating that peculiarly graceful attitude assumed
by the squirrels while sitting upon their hind legs feeding,
after imitating the curve of the back, — which not one in
a hundred can do naturally, — place the joints of the hind

legs so far up, and at such an acute angle, and the joints of the forelegs down at such an angle, that the two will almost touch. This rule should always be followed.

The preceding method may be applied when mounting all animals below the size of a Newfoundland dog. Larger animals are mounted in the following manner: Fill out the space occupied by the muscles of the head and legs in the manner already described. Procure five iron rods, with a shoulder cut at each end, upon which fit a cap (Plate IX. Fig. 2, B); on the extreme end have a thread cut with a nut to fit (A), — the distance between the nut and cap should be about an inch and a half. Cut a piece of plank, an inch and a half thick, about two thirds as long and wide as the body of the mammal to be mounted; bore five holes in it, as indicated in Fig. 1, A. Fasten one rod (8) firmly to the skull by drilling a hole through the top and placing the cap in the proper position. Screw the nut on well (14), and place the lower end of the rod in the hole in the plank prepared for it (11); fasten it firmly. Now stuff the neck out with hemp to the proper size. Drill a hole through the hoofs, or bottom of the feet, into the hollow of the bones (2, 4); force the rod (7, 7, 7, 7) up through this hole, through the stuffing of the legs, and fasten them into the plank (5, 6). Force a wire into the tail and clinch it firmly in the wood (15). By winding up grass or hemp, imitate the various sections of the body taken out, and place them in the proper positions (16), making allowance for the plank and rods. Or a better way is to take casts in plaster of the different parts and place them in the proper position.

Everything must be solid, to avoid sinkings and depressions in the skin. In this way the student can mount an animal of any size by increasing the size of the rods and plank. The ends of the rods must be fastened into a plank stand (10) by passing them through holes drilled in it (17, 17, 17, 17).

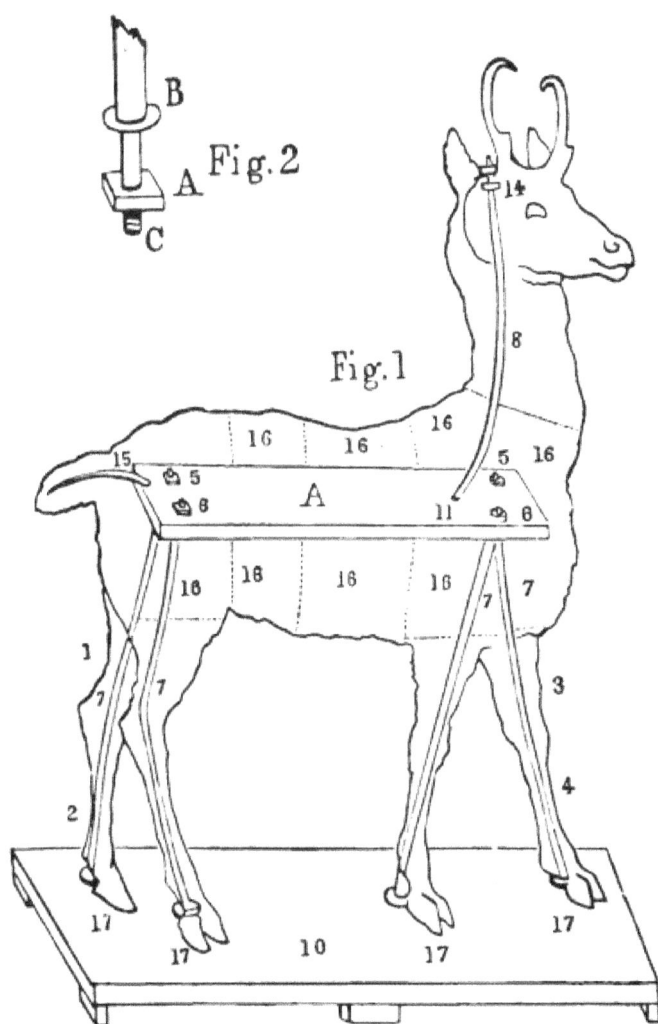

Plate IX.

To mount a dried skin, first soak it in alum-water until it is perfectly pliable, and then mount as before. The water should not be too strongly impregnated with alum, or it will crystallize upon the hair. About a quarter of a pound of alum to a gallon of water are the proper proportions. If the skull has been detached, replace it, or make an artificial one of grass or plaster to take its place. Mammals that have been preserved in alcohol may be skinned in the usual manner and mounted.

To skin mammals for the fur alone, cut in a straight line from the inside of the knee of one hind leg to the other. Skin as before explained, only cut off the feet and detach the skull. Stretch smoothly on a thin board, with the wrong side out. The skin should be lengthened rather than widened.

CHAPTER III.

THIS interesting class of animals has for a long time engaged the attention of students, yet it is surprising how comparatively little has been written about those of America. The almost infinite number of species still affords the young naturalist a wide field for careful investigation.

In collecting insects, the instruments used are : An insect-net, made of fine muslin or of silk gauze, and stretched upon a light steel wire frame, with a light handle, about four feet long, attached ; several wide-mouthed bottles and phials filled with strong alcohol ; insect-pins of the best quality, which can be procured at natural-history stores ; tweezers smaller than those used for birds (Plate I. Fig. 3); also, a small pair of pliers (Fig. 1) ; several soft-pine boards about twelve by twenty-four inches, planed perfectly smooth, will also be needed.

Boxes or drawers are necessary for the reception of the dried specimens, lined with thick felting or cork to receive the point of the pin that holds the insect and keeps it upright. An excellent box lined with paper is sometimes used to advantage, a description of which may be found in the "American Naturalist," Vol. I. p. 156.

I hardly need state that a good microscope is indispensable in prosecuting the study of insects, although it may be commenced without one. I shall take each order of insects separately, and endeavor to explain how they are collected and preserved, commencing with the

Beetles, or Coleoptera. — The best way to preserve beetles temporarily is by putting them instantly into strong alcohol; and as the collector will meet with specimens everywhere, he should never be without a phial ready for instant use. During spring and early summer thousands of minute species may be captured in the air with the net, especially just at night. During summer and autumn a great many nocturnal species may be captured near a light placed at an open window, or in the open air. Various species may be found feeding upon plants during the summer and autumn. A great many of the so-called carrion-beetles may be taken, during the same seasons, by exposing the carcass of an animal. Some species inhabit decayed wood, where diligent search should be made for them, especially in the woods, under old stumps or in them. Numbers of very beautiful beetles may be found in the excrements of animals, and under them, also under stones and logs of wood; they are found beneath the bark of trees and on sandy places, or in dusty roads. There are also a few aquatic species to be found in the water or near it.

To mount large beetles, force the pin through the right wing-covert near the thorax, and place the point in the cork, with the beetle's feet resting on it; place the feet in the attitude of life, with the antennæ in the proper position, with a pin on each side of them to keep them in place until dry. If the wings are to be extended, place the beetle on the pin as described; then, with an awl, bore a hole in the pine board; lay the insect upon its back, with the head of the pin in the hole; now open the wing-coverts, and spread the wings; over the latter lay a piece of card-board, and fasten it by placing pins through it into the wood on each side. The wing-coverts should not be fastened with a card, as it will flatten them. When dry, remove the card, and the wings will retain their position, when the beetle can be put in the proper position in the insect-box.

Smaller beetles, less than an eighth of an inch long,

should be fastened to a piece of mica or to a round bit of card-board with a little gum-arabic, and the pin placed through the mica or card, or they may be transfixed with very fine silver wire; this wire must then be inserted in a bit of cork, through which the common insect-pin is placed.

Beetles that are collected in remote countries should always be transported in alcohol. When they are to remain long in alcohol it should be changed once, then they will keep for years uninjured. After they have been in alcohol for two or three weeks there is no need of its covering them, as a little in the bottom of the bottle will keep them sufficiently moist; but they should never be allowed to dry.

Beetles may be preserved in a weak solution of carbolic acid as readily as in alcohol. This has the additional advantage of preserving the specimens that have been immersed in it from the ravages of noxious insects for some time. Glycerine can be used to advantage in preserving beetles that have delicate colors which fade in alcohol; but they cannot be pinned without cleansing.

Bugs, or *Hemiptera*, may be found generally upon plants. The common thistle (*Cirsium lanceolatum*) furnishes a pasture for several species. Numerous representatives of this order may be found on low bushes, and in the grass during summer and autumn. At least one species may be found in cheap boarding-houses during the midnight hours. The almost endless variety of Plant Lice come under this head, and may be taken everywhere on plants during summer and autumn.

These insects, like the beetle, are first immersed in alcohol, and afterwards placed upon pins, with the legs arranged in natural positions, and the peculiar sucking-tube, with which they are all provided, brought well forward so as to be more easily examined. The numerous

3*

aquatic species may be secured with a net ; they should be carefully handled, however, to avoid the sharp sting, or piercer, with which some of them are armed.

Grasshoppers, Crickets, etc., or Orthoptera. — Members of this order may be found everywhere, — the grasshoppers in the open fields and woods, where they may be caught in nets. The best way to kill them is to prick them on the under side of the thorax with the point of a quill that has been dipped in a solution of oxalic acid. If they are not to be mounted instantly, wrap them in paper. Crickets may be found in the ground in holes or burrows, under stones, and in the grass ; a few species may be taken on the leaves of trees or bushes ; some species of the well-known Cockroach may be found in houses, and some under stones and beneath the bark of trees.

All of the above may be mounted by placing the pin through the thorax, and arranging the legs as before described. The wings are also extended in the same manner as the beetles', with the exception of the wing-coverts, which are fastened with cards like the wings.

Walking-Sticks are found on low bushes or on trees, sometimes upon the ground. They are to be put into alcohol to kill them, then mounted like the beetles. These insects, when dry, require delicate manipulation while being moved, as they are *very* fragile. When the colors of the *Orthoptera* are to be preserved perfectly, place them in pure glycerine. This is especially necessary in preserving the larvæ of grasshoppers. Grasshoppers may be put into alcohol if convenient, but it must be very strong. This method will generally change the colors completely. Cockroaches and crickets should always be killed by placing them in strong alcohol.

Moths and Butterflies, or Lepidoptera. — All butterflies are diurnal, and are generally caught with the net. They may be killed by pinching the body just below the wings, or by pricking between the forelegs with the quill and oxalic acid used in killing grasshoppers. If they are not to be mounted instantly, they should be packed in pieces of paper doubled in a triangular shape, with the edges folded. Butterflies may be reared from the egg by capturing the impregnated female and confining her in a box pierced with holes to allow fresh air to enter. In this box she will deposit her eggs; these are allowed to hatch, and the larvæ fed upon the leaves that they naturally subsist upon. When sufficient time has expired they will cease to feed, and form a pupa or chrysalis, and either in a few weeks or the ensuing year come forth perfect insects, when they should be instantly killed. In this manner the collector will be able to secure fine specimens.

Although some few of the moths are diurnal in their habits, the greater part are strictly nocturnal. A great many specimens may be decoyed by the use of a bright light. During the months of May, June, July, August, and September, the following method may be practised with advantage in securing many specimens. Mix coarse brown sugar with alcohol enough to form a thick paste, saturate rags thoroughly with this paste, and hang them on trees or stakes in an open grove or wood at twilight; or daub some of the mixture upon the stakes or trees. This mixture, thus exposed, will attract the moths. The places should be visited every few minutes with a dark lantern, taking care not to throw the light upon the spot until near enough to catch the moths in the net if they should attempt to escape.

Mr. F. G. Sanborn — who informs me that he uses the strong-smelling New England molasses in the above-described manner with success — rightly remarks "that moths may be divided into three classes by certain species of them being affected differently by the appearance of artificial light in the night. One class are powerfully *attracted* by it; another class go about their usual avocations *unmindful* of it; while a third class are instantly *expelled* by it." The third class are by far the most difficult to capture.

Moths are easily reared from the eggs. In autumn and winter numerous cocoons may be found upon trees and bushes; these, if kept in a warm room, will hatch in early spring.

In mounting butterflies and moths I have practised the same method as described in mounting beetles, and think it superior to all others. In mounting these insects, however, it is well to use what is called a "setting-needle," to avoid rubbing the scales off the wings with the fingers.

The "setting-needle" is simply a common needle fastened into a light stick; two of these will be found useful, — one to hold the body of the insect firm, and the other to place the wings and antennæ in the proper position. The eggs and larvæ of the *Lepidoptera* should be placed in alcohol.

There is a class of moths called Hawk-Moths, Sphinxes, or Humming-Bees, some species of which are diurnal, and some nocturnal. They are all difficult to capture uninjured, as they fly rapidly, and, when caught in the net, struggle fiercely.

The larvæ, when about to form the pupa, go into the ground; for this reason the box that contains those that are being reared should be partly filled with moist earth. They are mounted in the same manner as the other *Lepidoptera*. All bright-colored insects when in the cabinet should be kept from the light as much as possible, especially those belonging to the above order.

Dragon-flies, etc., or Neuroptera. — Dragon-flies are, on account of their quick motions, somewhat difficult to capture; they are found flying over the fields and meadows; most abundant, however, in the immediate vicinity of bodies of fresh water. The lace-winged flies are also found in the vicinity of water. The larvæ of almost all of these insects are aquatic. They emerge from the water perfect insects. The larvæ should be preserved in alcohol. The perfect insects are killed with oxalic acid, and for transportation are packed in paper like the butterflies. When they are to be mounted, a copper wire is placed through the body and head; the wings are then spread, as before described.

Bees, Wasps, etc., or Hymenoptera. — Members of this order may be found everywhere in the fields and woods. Their larvæ generally resemble grubs, or maggots, and should be preserved in alcohol or glycerine.

The larvæ of the Ichneumon-Fly are found in the bodies of caterpillars. The larvæ of other species are found in the excrescences on various plants and trees. This class of insects may be caught in a net and placed in alcohol, or killed with oxalic acid. They are to be mounted as the other winged insects; the tongue must be brought forward so that it can be examined when the insect is dry.

The nests of the Wood-boring Bees, the Paper-making Wasps, and Hornets, the mud nests of the Mason Wasps, the excrescences on trees and plants, should all be collected and preserved dry after the larvæ has been taken out. Ants with their eggs and larvæ may be put into alcohol; it is best to capture these fierce little insects with the tweezers, to avoid their stings, which are sometimes poisonous.

Flies, Mosquitoes, etc., or Diptera. — These are the most difficult of all insects to preserve, especially when they have to be transported from a distance, as they must all be instantly pinned, with the exception of the Fleas, which may be put into alcohol.

They may be caught everywhere by beating bushes by the side of the roads and woods, then using the net.

Some of the species are nocturnal (as those who have slept in the open air in the woods during the warm months can bear painful testimony), and may be attracted by artificial light, as in the case of the moths, etc. Their larvæ are found in various situations, some being aquatic, others feeding upon putrid flesh and fish ; they may be preserved in alcohol.

In closing this chapter, I would impress upon the student the absolute necessity of labelling every specimen carefully, with the date and the locality in which it is found ; this may be done by numbers referring to a catalogue, as in birds and mammals, or upon a slip of paper. Also take notes of various circumstances relative to the habits observed at the time of capture, etc.

The best substance to protect cabinet specimens from the attacks of injurious insects is benzine, placed in an open vessel in each drawer or box. Camphor is also good, but I think that its fumes tend to fade the brighter colors of moths and butterflies. Spirits of turpentine is good, but it evaporates much quicker than benzine. Carbolic acid is, next to benzine, perhaps the best substance, if exposed in the same manner.

To mount insects that have been dried, place them in a box containing wet sand, and let them remain until soft, when they are mounted as before directed. I am informed by Mr. F. G. Sanborn that a few drops of carbolic acid mixed with the water used in moistening the sand will prevent mould from forming upon them while they are being softened. The same preventive might be put in the water used in moistening the cotton for softening birdskins.

CHAPTER IV.

COLLECTING AND PRESERVING FISHES AND REPTILES.

SECTION I. *Fishes.* — Very many are they who at the present day follow in the footsteps of the "Father of all Anglers," the good Izaak Walton, concerning the mere *sport* of angling; but, alas! there are few who, like him, look with contemplative minds upon the great works of Nature; for the worthy Izaak was quite a naturalist, after his fashion, and loved exceedingly to prate, in his quaint style, of the wondrous birds, beasts, and fishes of which he had seen or heard. Few, indeed, are they who, although some of their happiest moments are spent by the side of the clear mountain brook, with rod in hand, see in the beautiful trout, that they with exultation draw from its sparkling home, anything more than a good dinner on the morrow.

Yet there are a few earnest naturalists who love to study the finny tribes as they ought to be studied. Indeed, the science of Ichthyology can claim among its most earnest students the greatest naturalist in our land. Those who live inland do not possess the advantages of making as extensive a collection of fishes as those who reside upon the sea-shore; nevertheless, they can all do something for this branch of natural history.

In collecting fishes the instruments generally used are nets and hooks and lines; with these try and secure every variety that can be found. Many species can be secured from the markets, where fishes are exposed for sale, by picking out the specimens that are needed. The best way to preserve fishes is to put them into alcohol. All

large fishes should also be injected with alcohol before put-
ing them in it.

There is, however, another method by which fishes may bo
preserved; that is, by skinning and stuffing. Thus: Open
the fish on the under side from the throat nearly to the end
of the body, or within a short distance of the root of the tail;
then skin down each way, taking care not to scrape off any
of the pigment that covers the inside of the skin and gives
the fish its color; cut off the fins close to the skin on the
inside, also the head at the gills; clean out the brains by
enlarging the hole in the occiput, where the spinal cord
enters the skull; remove the eye from the outside, dust
arsenic into the orifice left, and fill it with cotton; cover
the inside of the skin with arsenic; fill it to the natural
size with cotton, and sew it up; place a wire transversely
through the fins to keep them in position.

Another method is to remove the skin from one side,
and to clean the flesh out in this way; the fish is then
stuffed and placed upon its side, so that the opening will
not show. This method will answer very well for flat
fishes, but large ones must always be stuffed in the manner first described.

Section II. *Reptiles.* — Many a harmless snake or toad
has been sacrificed to ignorance and superstition. Indeed,
so strong is the general prejudice against the most common snakes, — which are as incapable of inflicting an injury
as a mouse, — that but few persons will hesitate to kill the
supposed venomous reptile at sight, if indeed they have
the courage to remain long enough in its vicinity to do so
valiant a deed. Such persons really believe that they are
removing a dangerous adversary of man from the face of
the earth. I would, however, advise them to glance for a
single instant at the history of these interesting — although, I will allow, somewhat disgusting-looking — animals before they again shed innocent blood. All the snakes

in Massachusetts may be handled with impunity, with the exception of two species, which are very rare. I refer to the Copperhead and Rattlesnake. The prettily marked Milk Snake, or Checkered Adder, and the imaginary terrible Water Snake, are quite harmless, although we are everywhere informed by those who are ignorant upon this subject that they are exceedingly venomous. So long as people are erroneously educated in this belief, so long will the poor snakes suffer unjustly. Snakes, with but few exceptions, are neutral regarding the interest of man.

The best method of preserving snakes is to put them into alcohol moderately strong, as otherwise the scales start easily. Snakes may be benumbed by thrusting a pin into their brains; in this way they may be carried from place to place more readily than if they were uninjured.

Snakes may be skinned after making a longitudinal incision, about two inches long, in the largest part of the body, on the belly; then by drawing back the skin, the body may be divided, and the parts drawn out each way. The head should not be skinned. The eyes are removed, as in the fishes, from the outside. The skin is now covered with arsenic and turned back. It is then filled with bran to the natural size. It may, after sewing up the incision, be placed in any position desired. Artificial eyes are fixed in the head.

If the head is to be raised, run a sharpened wire through the top of it, and through that section of the neck and body that is to be elevated, through the skin into a board, cut off the protruding end, and close the skin of the head over it. After the skin becomes dry, the wire can be taken out of the board, and cut off close to the body.

Turtles may be preserved in alcohol, or they may be skinned and mounted thus: With a small steel saw cut out a square section on the under shell; remove this and draw the intestines, bones, and flesh of the legs, etc., out of the hole thus formed; skin the legs down to the toe-nails, removing everything; skin the head and neck; cover the inside of the shell and skin with arsenic. Turn the feet and neck back, and stuff them to the natural size with cotton. Fill the neck with bran; roll up a small ball of grass, place it inside of the shell; then force a piece of wire through it into the head, and clinch the end in the ball. Pack cotton or hemp around the grass in the shell, to keep it firm, and to fill up the empty space; then re-place the piece of shell taken out, and fasten it with glue or putty.

Now put the animal in the proper attitude upon a piece of board, and arrange the feet in the natural position, and pin them until dry; place the head naturally. The eyes should be removed from the outside, and artificial ones substituted. If it is not convenient to skin a turtle, place it in boiling water a few moments, when the softer parts can easily be removed from the shell. In this case, how-ever, the bones and skull should be cleaned, labelled, and preserved with the shell.

For scientific specimens, toads and frogs must be pre-served in alcohol. But they may be skinned in the follow-ing manner: Open the mouth as wide as possible, and cut through the bone of the neck or back from the inside; do not cut the skin; then separate the flesh on the inside all around. Take hold with the thumb and forefinger, or with a pair of pliers, of the backbone, and press the skin downwards, and draw the body out. When the forelegs appear, cut the bone and flesh off to the toe-nails, and pro-ceed to perform the same operation with the hind legs. Cover the skin with arsenic, and turn it back, — the legs may be easily turned by blowing into them with the breath. Fill the body with bran, and support the head in a natural position with cotton until dry. Remove the eyes from the outside, and supply their place with artificial ones, but be sure to place them in the proper position. To place a frog or toad in a fancy attitude, place a ball of grass in the body, and wire the legs as described in small mammals.

The best time to collect toads and frogs is during the breeding-season in spring. The salamanders may be found under stones and logs in damp places; also some species in springs and clear running brooks, under stones. They must be placed in alcohol at once.

Lizards and alligators may be skinned in the following manner: Make an incision the whole length of the belly, and skin as described in mammals, leaving the skull in. Do not try to remove the skin from the top of the head, as it will be likely to tear. The leg-bones should be cleaned and left in. The reptile is then mounted in the same manner as a mammal. Lizards and small alligators may be put in alcohol.

The eggs of frogs and of salamanders may be preserved in alcohol. The eggs of lizards, alligators, and turtles may be blown in the same manner as birds' eggs; but it is well to place some in alcohol if they are in an advanced state of incubation, as they will serve to illustrate the growth of the embryo. But the egg must be broken slightly to admit the alcohol to the embryo.

Last winter I accidentally made a discovery relative to the preservation of fish and reptiles. While travelling in Florida, I accidentally lost some alcohol. Being unable to replace it, and having some reptiles to preserve, I put about an ounce of carbolic acid into a glass jar, with half a pound of arsenic; to this I added a quart of water, — I will here remark that the waters of Florida are strongly impregnated with lime. Into this composition I put some reptiles and a few young mammals. After two weeks, the jar was packed with others in a box, and sent North by express.

Upon arriving home, and opening the box, I found that the jar had become broken, and the liquid had escaped. The smaller reptiles, etc. I placed in alcohol; but a reptile known as the "Glass Snake" and a young Rabbit were left out for want of room, set away and forgotten. Upon looking them up about a month afterwards, I found, to my surprise, that the "snake" *had dried completely without shrinking in the least,* and, moreover, it *retained all the peculiar glossiness of life!* The Rabbit had not shrunk any more than if it had been in strong alcohol.

Such is the result of an accident. Whether this discovery will prove of general practical use in preserving reptiles is yet to be proven.

CHAPTER V.

SECTION I. *Crustacea.* — But few of these interesting objects of natural history live away from the salt water. The Crawfishes and a few others form the exceptions to the rule. All Lobsters, Crabs, Shrimps, and Crawfishes may be preserved dry. Wash them in fresh water, and, if the specimen is large, remove the flesh as much as possible by lifting the shield, or upper part of the shell. The specimens should be placed in as natural an attitude as possible to dry. When dry they should be handled with care, as they break easily. If arsenic is put into the body, it will help to preserve it and keep away noxious insects.

Small Crabs, Shrimps, etc. should be injected with carbolic acid and dried carefully. Never place a specimen in the sun to dry, but always in a draught of air in the shade. A great many kinds of Shrimps or Sand-Fleas may be collected from under sea-weeds on sandy beaches.

Collecting Mollusks. — Many shells may be collected on the sea-shore among the rocks at low tide. Some of the more minute species may be found clinging to the sea-weed that grows on the rocks. These require delicate manipulation, as they are very fragile ; they are best removed with the tweezers, and should be placed in wide-mouthed bottles containing alcohol. Some species of cone-shaped, univalve shells may be found clinging closely to the rocks. They should be seized suddenly with the hand, and, before the animal has time to contract itself, — which it will do very quickly, and then it adheres so closely as to render its separation from the rock without injuring the shell extremely difficult, — removed with a sliding motion.

Many species may be found buried in the mud and sand below high-water mark. The exact locality where these are hidden may be determined by searching for their breathing-holes on the surface of the mud or sand ; then, by carefully removing a few inches of the soil, the shell may be detected. Numerous species may be taken in deep water by dredging, or with a rake, such as is used in gathering oysters, etc.

A great many shells may be procured just as they are cast on shore from the action of the waves ; these must be washed in fresh water and dried. The different species of smaller fresh-water shells may be found upon rocks, aquatic plants, and on the surface of the mud. They should be placed in alcohol. The larger species — such as the mussels — may be taken by dredging. Numerous shells of mussels may be found at the entrances of the holes of the muskrats ; of these the collector may take his choice, as many of them are in excellent condition for the cabinet.

The land shells, or snails, may be taken from the different plants upon which they feed, or from under stones or logs, especially in damp places. The smaller species should be carefully removed with tweezers, as they are very fragile, and placed in alcohol.

Preserving Shells. — It is well to preserve in alcohol numbers of all species of shells containing the animal. To remove the contents from shells that are to be dried for the cabinet, boil them a few moments, and clean them with a bent pin or wire. The contents of the different species of bivalves may be removed with a knife without boiling, as by this method the shell retains its color much better. The bivalves should have their shells closed and tied until dry. If the shells of mussels have a chalky appearance, it may be removed by immersing the specimen for a few moments in a bath of diluted muriatic acid. All shells should be carefully washed in fresh water with a tooth-brush.

Never varnish a shell; it shows bad taste to try to improve upon nature in this way, besides injuring the specimen for scientific use. As some of the more fragile land shells are liable to crack when drying, it is well to apply a slight coating of gum-arabic dissolved in water. This at some future time may be easily removed. There are also some species from which the epidermis is liable to peel; to prevent this, Mr. F. W. Putnam informs me that they should be immersed in oil for a short time.

Worms. — Marine worms may be found in the sand or mud and under stones. They should be kept in strong alcohol. Earthworms, Leeches, etc. must also be kept in alcohol.

Many species of marine worms may be found in the hulls of ships, or in wood that has been immersed in salt water for some time.

Animal Parasites. — Recently in this country, and for some time in Europe, attention has been directed by eminent naturalists to the parasites found on birds and other animals, and in their intestines. These should be placed in alcohol. The parasites from each bird or animal should be kept separate, in small phials, with the name of the bird or animal from which it was taken attached, also the date and locality.

The *Jelly-Fishes* may be found in deep water or near the shore in countless numbers. There are a great many species. They may be preserved in the following manner: After catching them in a bucket, pour off the water, and add strong alcohol, a little at a time. The animal will give out water continually during this operation, and alcohol

should be added until it dies, when the water will cease
flowing. It should then be removed from this solution
and placed in strong alcohol, where it must be kept
permanently.

Corals — which generally grow at some distance from
the shore, and sometimes in deep water — should be secured with nets. They must first be washed in fresh water,
then dried in the shade. It is also desirable to preserve
specimens in alcohol.

Sea-Anemones are found attached to the rocks or buried
in the mud; they should be plunged in strong alcohol
when fully expanded, but the alcohol should afterwards be
changed, as they give out large quantities of water.

Hydroids and Bryozoa. — Incrustations on the rocks, sea-
weeds, and delicate tufts found growing on rocks, etc.,
are called by these names. They may be dried or preserved in alcohol like the Corals.

Star-Fishes may be found among the rocks at low tide.
They should be killed by immersing in alcohol or fresh
water. Some species should be preserved in alcohol, where
they should be placed in as natural attitudes as possible, as when they become rigid it is impossible to alter
the position of the arms. They may be dried *in the
shade* by placing them in natural positions upon a board.
When dead, they should be dried instantly, as they will
decompose in a few hours if kept in a damp place.

Sea-Urchins may be taken in rocky pools at low water.
They may also be found under the sand on beaches, from
which they are frequently washed by the waves. They may
be preserved in alcohol, or dried like the Star-Fishes.

Holothurias, or *Sea-Cucumbers*, are found on flats or
under stones. They must be preserved in alcohol.

Sponges and *Seaweeds* should be dried in a draught. Very
pretty ornaments are made of the sea-mosses by washing
them in fresh water, and spreading upon dampened paper

with a fine needle; the glutinous matter contained in the plants will cause them to adhere so firmly to the paper when dried and pressed as to look like a very fine engraving or painting. When a collection of these are executed by a skilful and artistic hand, and bound in a book, they form a beautiful and interesting volume.*

SECTION II. *Preparing Skeletons.* — I will give the methods by which bones may be cleaned. To clean the bones of large animals, first take off as much of the flesh as is possible with a knife; then put them in slatted boxes, and place the boxes in a running stream, or between tide-marks on the sea-shore. The boxes, being open, will allow the entrance of Shrimps, other aquatic animals, and insects, who will devour the meat, while the water, having free passage through, will perform its part. When well cleaned, wash them in warm soap-suds, and, after rinsing, dry in the sun and air; this will tend to bleach them.

The bones of smaller animals may also be cleansed in this manner; but the better way is either to boil them until the flesh comes off easily, or to put them into water that has been impregnated with chloride of lime; in both cases the bones will have to be cleaned afterwards with a knife and a stiff brush; they should be scraped as little as possible. If kept in a dry place, exposed to the action of the air, the bones will bleach constantly.

Mounting Skeletons. — To mount the skeleton of a bird, place a wire through the hole occupied by the spinal cord, and fasten it in the skull; this will hold the vertebra of

* As there is not a general interest manifested in the objects alluded to in this section, I have given but few directions for collecting and preserving them, but such as will, perhaps, satisfy the general collector. Those who are particularly interested in them will find in the pages of the various numbers of the "American Naturalist" more particular directions for collecting and preserving each branch of this truly interesting class of animals, written by the most competent and well-informed men in our country

4

the neck and tail, and other bones of the back, in position. Next, force a wire through the hollows in the bones of the tarsi, tibia, and hips (Plate X. k, y, j) by drilling a hole through each end; now fasten this wire to the broad bone that covers the back (m), by drilling a hole through on each side and bending the wire down firmly (x), first over then under the bone, where it meets the end of the opposite wire; twist the ends together. The wing, breast, and other bones are now fastened on by drilling holes transversely through the ends and running wires through and twisting them (r, d).

The skeletons of mammals, fishes, etc. are mounted in much the same manner. If large, they are supported on iron rods. The wire used must be composed of brass or copper, as iron corrodes easily. The fleshy or cartilaginous parts of the feet should be removed, but not the outer or horny portion of the bill.

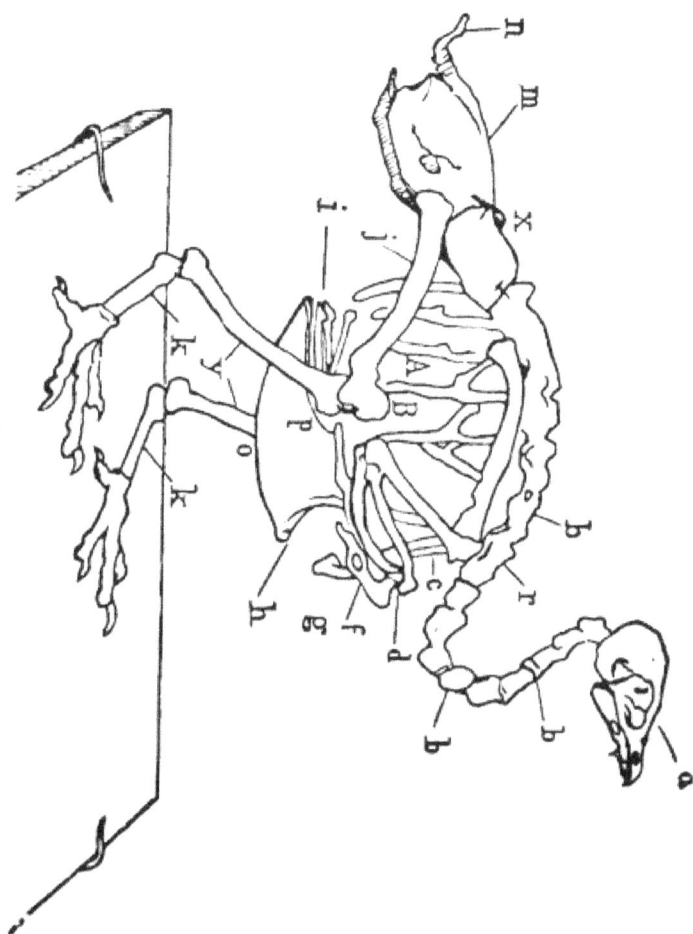

Plate X.

CHAPTER VI.

COLLECTING AND PRESERVING EGGS.

No portion of natural history has received more attention than the science of Oölogy; yet in very many cases collections of eggs are made in such a careless manner as to render them worthless, except as ornaments, on account of the collector's not paying sufficient attention to *identification* and *authentication*.

Let identification, then, be the collector's first care; let him make it a rule *never* to take an egg or nest until he can surely tell to what species it belongs. The best method of learning the name of the owner of the nest is to shoot her, especially by collectors who have had but little experience in studying birds; while the more practised ornithologist can generally tell at a glance, if the bird is large, what it is. While collecting the eggs of the Warblers and other small birds, the most experienced oölogist should *never* neglect to shoot the bird, even if he has to watch for it a long time.

Nests and eggs should never be labelled on the authority of a person who has found them, and only *seen* the birds, but who is in a comparative degree unacquainted with them. The nest should be seen *in situ*, and the bird identified. I have known a great many errors to arise from this source.

Commence early in spring to look for the nests of the rapacious birds, and continue the search for these and other nests until late in summer. I know of no rule to be followed in finding nests. Search long and diligently in every locality frequented by birds; and watch them while building. Place straw, hay, cotton, hemp, or any of the materials that birds use in constructing their nests, in an exposed situation in a swamp or wood, then by watching the birds when they come to take it, and following them, many nests will be found that would otherwise escape notice.

' To remove the contents of an egg, drill a small hole in one side with a drill made for this purpose (Plate I. Figs. 5, 6) ; two sizes of these drills will be required. Now, with the blow-pipe — of which two sizes are also needed, (Fig. 7) — applied to the lips, force a small stream of air into the hole ; this will cause the contents, if fresh, to escape at the *one* hole. To prevent breakage while drilling the eggs of the Humming-Birds, or other small birds, it is well to cover the outer surface with thin paper, gummed securely on, and dried.

To remove the contents of an egg that has the embryo partially developed, drill as before, only a larger hole is necessary ; then with a small hook (Fig. 8) remove the embryo in small pieces ; after which introduce water with the blow-pipe to rinse the interior of the egg. If the contents are allowed to remain in a few days, it will facilitate their removal. If the egg is covered with paper, as in the case of the Humming-Birds, the edges of the hole will be less liable to be injured by the shell being broken while using the hook.

Never make holes at the end of the egg, or on opposite sides , but if this old method is still preferred, they should both be made *on one side*, with the larger one nearest the greater end.

The best method that I know of for authenticating eggs is the following : After the egg is blown, place a number, written with ink, upon it, corresponding with one placed in the nest, then draw a line beneath it ; under this line place the number of the egg in the nest : thus $\frac{29}{4}$ would

mean that the nest is No. 29, and the egg is the No. 4 of
that nest; both of these numbers will refer to a book,
where all the particulars of the finding of the nest, the
locality, measurements of the nest, eggs, etc. in inches,
are recorded.

The method of preparing a book like that referred to
above may be seen in the following specimen : —

Scops asio.

Remarks. No.	Diameter and Depth of Hole.	Situa- tion of Nest. Height from Ground.	Tree.	Land.	Date.	Locality.	No. of Eggs.				
							5	4	3	2	1
29 Nest com- posed of leaves	5 In. 12 In.	30 ft. Hole	Oak	Low	1868. April 27	Weston	Length 1.50 Width 1.25	1.47 1.27	1.45 1.25	1.46 1.20	1.50 1.27

The measurements of an egg are taken with the dividers in hundredths of an inch. The number is attached to the nest. Nests, if composed of loose materials, must be kept in boxes, separated from each other; if lined with feathers, benzine should frequently be applied, to prevent their being attacked by moths.

APPENDIX.

CHAPTER I.

COLLECTING AND PRESERVING BIRDS.

SECTION I. *How to collect.*— First, let me say a few words to the young Naturalist. When I last appeared before him as an adviser, I rather intimated, that members of our brotherhood were looked upon by the world at large as slightly insane, or at best, as very foolish. This was much more the case then, than it is to-day. *Now*, the tables are turned, as it were; the scientist need no longer dread the scoffs of the "practical men," for the knowledge which he possesses is the "open sesame" for him to all ranks of society. The tide of popular opinion now flows strongly towards the gate of learning opened by modern zoölogical science, and all are eager to listen to what may fall from the lips of our eminent professors.

Thus, many obstacles which were formerly thrown in the path of the young and enthusiastic collector, are removed. A word to parents or guardians, however, may not come amiss just here. I have often been asked, "Do you think the study of Natural History will prove beneficial to my son?" My answer is, "Most certainly it is beneficial; but, aside from direct benefits, which are, perhaps, too numerous to mention, there is one important point which ought to be kept in mind. While your son is engaged in this enchanting study, his mind will be so fully occupied with the

81

multiplicity of objects which are constantly inviting his earnest attention, that he will find but little time to devote to mischief. Young and active brains *will* find occupation; the old hymn wisely says, 'Satan finds some mischief still for idle hands to do.' Leaving his Satanic Majesty out of the question, it is best, as all will agree, to allow our children to occupy themselves, during leisure hours, in some amusing recreation. What can be better than the enjoyable and health-giving exercise connected with the study of Natural History?"

So much for moralizing; now for the subject. I want to reiterate what I have said in the latter portion of the third paragraph, page 4. No matter *how common* a desirable species may be, when you meet with it in a locality with which you are unacquainted, proceed at once to collect all you want. I have known of many instances where delay, in such cases, was dangerous, the birds having disappeared in a single night.

I have somewhere seen a remark made by a distinguished Naturalist to the effect, that any one could collect birds after a short experience. This I deny, and will further state, that not one person in fifty will ever make a good collector, for the gathering together of birds is a high art. In order to become an eminently successful collector, many acquirements are necessary. A quick eye, a good ear, perfect coolness, accompanied with ready action in emergency, patience in an extreme degree, a tenacious memory, and an utter disregard of such minor troubles as wet feet, scratches from thorns, stings from insects, etc., are among the requisites in the make-up of a first-class collector.

But the tyro must not become discouraged by reading this array, for many of these attributes, if not all

of them, can be acquired by practice. Study well the
habits of each and every species which comes under
your eye. Note carefully song or motion, and learn to
even distinguish the various Sparrows and Warblers by
the chirp alone. Although this is somewhat difficult,
it can be done, for I know of several who do it.
No two species possess habits which are exactly alike
in every respect, and the nicer points of distinction
can be learned so that the flirt of a tail or the droop
of a wing will often betray a rare bird, even if it be sur-
rounded by hundreds of more common species which
are quite similar in appearance.

This art may be carried to such perfection, that it
appears like something marvelous, to one who is not
skilled in it, to see how readily certain obscurely-col-
ored species may be detected, even at a long distance,
from among others having similar form and markings.
I have frequently known two experts, when collecting
in company, to shoot, both at one instant, at the same
bird the moment it appeared, although neither was
aware that the other saw it. By learning to distinguish
all species instantly, the ornithologist is spared the
trouble and pain of shooting birds which are too common to be of any value to him.

In addition to the young, in all stages of plumage,
moulting birds should also be taken, as many valuable
facts can be learned by studying the various changes
undergone at this stage.

More recent improvements in breach-loading guns,
now render them desirable ; in fact, a muzzle-loader is
scarcely to be thought of for a collector. I do not
here recommend any particular make, but will simply
state, that I am at present using a Parker gun, and
find that it works like a charm. In collecting little
birds, however, I use a gun of an exceedingly small

calibre, thirty-eight hundredths of an inch being large enough. With a proper charge, which can be ascertained by experiment, using equal bulk of powder and shot, birds as large as Blue-birds can be killed at twenty yards. Besides being more economical, the specimens collected with this gun are better shot than with a heavier one. The report is lighter, and does not frighten the birds as much.

I do not now recommend Ely's wire cartridges. For large birds, use a rifle with a small calibre. This will kill farther, and will usually insure good specimens.

In addition to the advice given on page 7, first paragraph, I would remark, that unless the shot-holes in the abdomen are found and carefully plugged, the escaping fluids, being often exceedingly acrid, are quite apt to soften the skin in a short time. I have known the epidermis to slip on the abdomen before the specimen was cold, from this cause.

In picking up a Heron, Duck, or Wader which has fallen into muddy water or ooze, care should be used to take it by the bill, as then a greater portion of the filth will slide off the oily feathers, which process may be facilitated by gently shaking the bird. I have seen white herons completely ruined by collectors who took them out of the mud by the feet, thereby allowing the dirt to slip under the immaculate feathers.

Allow me once more to repeat the caution about handling guns. A good breach-loader, if properly handled, is perfectly safe: but never point your gun at a human being (there is no necessity of getting in front of the muzzle yourself when it is loaded), and there is no danger to any one. In shooting, the gun should never be brought to a full cock until you are about to fire. Practice will make perfect in this respect. I can

cock my gun, even when Snipe shooting, after the bird
rises, and kill; further, I can shoot two Quail which
both jump at once and fly in opposite directions, cock-
ing each barrel for each bird after they are a-wing.
Almost any one can do this with practice.

Birdlime may be used to advantage in securing
birds, especially during the breeding season. A small
twig is covered with a thin layer of this exceedingly
viscid substance, and placed in such a position that
the bird will alight on it when she goes to the nest.
The limed twig should be lightly poised, so that it will
fall a short distance, as this will cause the bird to
stretch out its wings, and thus become entangled more
firmly. The lime should be spread with the fingers,
which should be first wet, to prevent its sticking to
them. I have also used various traps, snares, etc., to
advantage in capturing birds. The blow-gun is also
very good, although it is somewhat uncertain, as pro-
jectiles fired from it are apt to glance from twigs,
leaves, etc.

SECTION II. *How to prepare Specimens, Instruments, Materials, etc.*

To the instruments mentioned add three or four
sizes of awls, made long for boring the feet and tarsi
of dried skins.

I have now given up the use of arsenic, as being
very poisonous, for I am convinced that my health
has suffered from using it. I have, however, discovered another substance which I think far superior to
arsenic. This is a product of coal-tar, and resembles
carbolic acid in its effect as a preservative. The odor
is also disagreeable to insects, and specimens preserved
with it are free from their attacks. This new preservative is in the form of a powder, and is not a dangerous poison.

I will once again refer to the poisonous gases engendered by birds in progress of decay. I have been surprised to learn from competent physicians, that little or nothing is known of the baneful effects of this gas. Several years ago, when suffering from the effects of it, I visited an old physician, and described the symptoms to him. I was then ignorant of the cause of my illness, so was not surprised when he informed me that I had been poisoned with ivy (*Rhus toxicodendron*). His treatment of the difficulty, while laboring under this mistake, had little or no effect. I then consulted another doctor, who, although well acquainted with my pursuits, also decided that I was poisoned with some vegetable substance. His treatment also failing, and being then convinced that both were wrong, I discovered the cause for myself, and the remedy, which is as stated on page 14.

I now recommend the wide, thick shoes called army shoes for collectors; they are doubtless, the best in summer; in winter either leather or rubber boots, according to the season.

I now use forms slightly different from those mentioned on page 18 ; these are strips of tin, rolled up in half-cylinders, resembling a single section of those figured in Plate IX., Figs. 1 and 2.

SECTION III. *Measuring, Skinning, and Preserving Birds.*— In skinning small birds, time may be saved by breaking off the end of the tibia; then, by stripping downwards and twisting, the muscles may be all cut at once. The brains may be removed much more easily by three cuts; one down through the base of the skull, as described, and two on each side, beneath the skull; these last meet under the eye sockets; thus a triangular piece is removed, to which the brains adhere.

In large birds, like Eagles, I now skin over the metacarpus (beyond the carpal joint). This may readily be accomplished with practice. Indeed, every bone in the wing, including the phalanges, may be removed from the inside. I have, also, frequently performed the somewhat difficult feat of removing every bone in the body, including those of the bill and claws, leaving only the horny,covering ; thus securing a perfect skeleton, as well as a mounted specimen.

I do not now tie the bones, as mentioned on page 23, but simply place the wings in position.

When placing the cotton in the neck, as described on page 24, be sure that the end of the roll enters the cavity of the skull; this will make the neck more solid.

I now sew through a pinch of skin from the outside, when fastening the wings in position, by the sides and *over* the quill, not through it. Thus the thread is tied *outside*.

In filling small birds, I now first sew the wings, and then place neck and body in together. This saves time, and makes a better skin, it being stronger.

Too much care cannot be exercised in placing the skin in the form. To make a perfect specimen, every feather should be carefully put in place, and smoothly arranged. For drying, place the skin in a closet, or some place where it will not be disturbed by the slightest breath of wind. I now recommend writing date, locality, and sex, on labels attached to the skin ; also in case of rare birds, color of feet, bill, eyes, etc.

In preparing the wings of large birds, like Hawks, first fill the neck and body, using grass or excelsior for the latter. Then, after placing the wing in position without drawing the forearm within the skin, sew at the sides as in small birds; and also at the carpal joints.

Ducks are treated in the same manner; but always turn the head on the back, and stitch the feet together. Open on the back of the head, instead of on the throat, when skinning.

I do not now bend the necks of Herons, but simply lay the head on the back, stitch the legs together at the tarsal joint, bend the legs forward outwardly, then fasten the toes to the wing. This gives the skin a compact form.

Some three or four summers since, I was accidentally left on a small islet lying in the midst of the Gulf of St. Lawrence. As this lonely rock was swarming with birds, all of which were desirable, I soon collected a large quantity; but, unfortunately, when I visited the rock I only intended remaining a few hours, so did not go provided with arsenic for making skins. This omission, although I then considered it a misfortune, proved of great benefit to me, as it resulted in a discovery which has since become invaluable. Having skinned a large quantity of birds, and as the vessel which was to take me off did not arrive, I was at loss to know what to do with them. But, as necessity is the mother of invention, it occurred to me that I might salt them, and thus carry them home. There was a light-house on the rock, and the keeper had a supply of salt; I procured some, and rubbed it on the skins. These were afterwards simply packed in barrels, and sent to Massachusetts. When I came to use them, I found them in perfect condition, only requiring to be washed, when they came out like fresh skins.

I have since applied this method to all large skins, and find that it proves effective, even in Florida. The skin is removed as usual, and simply salted; the salt

being applied as if it were arsenic or any other preservative. The skin is then folded neatly, and wrapped
in paper. When ready for use, it is put into the damp-
ing box for a day or two; then the inside is carefully washed, the preservative is applied, and the
bird is mounted.

SECTION IV. *Mounting Specimens.*— I do not now
fill the neck with any loose substance. The body is
made as before; then a wire is pushed through it
lengthwise, firmly clinched behind, and protruding in
front as long as the neck, skull, and one half the bill·
This is wound with hemp or cotton as far as the skull,
care being taken to make it somewhat smaller than
the neck. This is coated with a layer of clay, mixed
to the consistency of putty, well kneaded, making it
the size of the natural neck, excepting that it should be
larger at the base. The brain and eye cavities, as well
as the space occupied by the tongue, should also be
filled with clay; by using this pliable substance, the
neck and head can be placed in any position. Well-
kneaded clay becomes as hard as stone, when dry; it
also possesses the advantage of not shrinking. Clay is
especially useful in mounting dried skins. I also fill
the tibiæ of Hawks, Herons, etc., with it.

Instead of pinning up the wings, as described on
page 39, they should be wired. Cut wires of a suitable
length, of a smaller size than is used for the legs, and
pass them through the wings, entering them just below
the carpal joint, on the under side; thus on through
the body, clinching as described for the legs. The
outer end of the wire is now passed through a small
opening which occurs in the carpus (seen near F, Plate
X.), brought out above, and bent firmly down. This
wire will always be concealed from above by the spurious wing which lies over it.

Care should be taken to find this opening in the carpus, for if it be passed between the phalanges it will separate them and the quills.

In raising the wings, this method of wiring will be found of great advantage; the secondaries, etc., should, however, be kept in place by supplementary wires, until dry, as before described. In large birds, supply the place of the wing muscles with clay.

I do not now recommend opening Ducks, or any other birds, under the wings; but if any grease remains on the skin, coat it with soapstone dust, and it will never give trouble.

NOTE TO SEC. II. Another instrument which I now consider necessary in mounting and making skins is a scissor-like tool, having long flat blades or points, called a stuffer, and which may be obtained of almost any dealer in naturalists' supplies.

I have now not only given up the use of arsenic but have also abandoned the use of napthaleine, the product obtained from coal tar, of which I speak in the appendix to this chapter. In place of this I have succeeded in manufacturing a Preservative of several ingredients, which, besides not being a poison, is a deodorizer, completely absorbs oil from greasy skins, preserves them better than arsenic, and is equally good in preventing insect attacks.

As will be seen upon referring to the next section, a new method of skin-making renders the use of tin forms unnecessary, or, in fact, any other forms.

NOTE TO SEC. III. *A New Method of Skin-Making.*
I have, in the last few years, considerably changed my method of making skins, and now proceed as follows:

The skin is removed as directed, but before it is turned a piece of wire, varying in thickness according to the size of the bird, is twisted around each wing bone, connecting them together, but they should be kept as far apart as they were when attached to the body. The skin is now turned as before; no sewing is now done to the wings, as the wire will keep them in place. Place the cotton in the skin in one piece, as directed, taking care that the wing bones lie parallel with the body, for if they cross one another the wings will not set well. Sew up the orifice, smooth the feathers, and see that the wings lie according to directions given previously. A piece of cotton sufficiently large to envelope the skin is now split into very thin layers, the skin placed on one of these, and wrapped in it by drawing first one side, then the other, over the bird, thus covering every part, even the head. Skins prepared in this way are simply laid one side upon any level surface until dry, when the wrapping may be removed; or, if the skin is to be packed for transportation, a thicker layer of cotton is added. The cotton which is used for this purpose is of a nice grade, and is advertised in our supply catalogue.

Large birds are treated in the same manner, and the necks as well as the legs of all birds should be kept straight, in as natural a position as is possible. The skins of ducks should be placed on their breasts, as more characteristic colors are to be seen above.

Salted skins should not be kept for more than a year without making over, or mounting, and the feathers should be kept as smooth as possible.

NOTE TO SEC. IV. Of all the methods of mounting which I have practised, I prefer the one now to be described: No hard body is made, but the cotton is wound around the neck wire as mentioned in the appendix; placed in the skin, leaving the lower end protruding as far back as the root of the tail. Wires are fastened in the wings as directed in the preceding instructions, and the ends which protrude into the body are wrapped firmly about the neck wire; next the leg wires are pushed in and also wrapped around the other wires; after which a tail wire is also fastened on. The space around these wires is now packed with cotton to the natural size of the bird, care being used to place the cotton in layers, not in bunches; sew up the orifice, mould the bird into form somewhat, and place it on a stand, where it can be finished as previously directed. In using this method, smaller wires can be taken, as the body is not as heavy as when excelsior is used.

CHAPTER II.

SECTION I. *Collecting.*— A good way to capture small Mammals in an unsettled section of the country, is, to dig a pit, which may be partly filled with water. A great many of the smaller Rodents, as well as Shrews, Moles, etc., will fall into this during the night. I have practiced this with success while in Florida.

SECTION II. *Skinning.*— I now say that the tails of both skunks and musk-rats may be skinned in the ordinary way : *i. e.*, stripped out.

My method of skinning Mammals has changed somewhat. Only small Mammals are to be skinned as described, and the following exceptions are to be made : Do not leave any bones in the skin; I even remove the bones of the claws, if I wish to mount the skeleton. This can easily be accomplished, with practice, for the horny covering of the claws will come off readily, especially if they be split on the under side. Always remove the skull.

To make a skin of small Mammals, fill it with cotton, and proceed as directed. Label the skull, or, what is better, attach it to the skin. This filling should only be used when a permanent skin is needed for the cabinet. If it is to be mounted at some future time, simply coat the skin with salt, and either wrap it in paper, or pack in a box by itself.

In skinning large Mammals, the crosscut should be continued to the foot, and all the bones removed.

93

SECTION III. *Mounting Mammals.*—There is no part of this work wherein I have made so much improvement as in the present section.

For many years I have endeavored to find some method by which Mammals could be mounted, and still retain the life-like fullness of the muscles, especially those in the region of the head. In order to accomplish this, I knew that I must fill the parts with some substance which would not shrink upon drying, and yet be readily moulded. Happily I have found two materials which admirably answer the purpose, viz., clay and plaster.

In mounting very small Mammals, either supply the place of the muscles of the skull with plaster, or make a cast of the entire head, taking care in both cases to insert a wire so that it shall protrude out of the back of the occiput. Fill the legs with clay, place the skull or cast in position, then fill the neck with clay, and proceed as before directed, only using clay in place of bran. It must be remembered that Mammal mounting is extremely difficult, and that it requires long practice to acquire anything like perfection ; yet, if complete measurements have been taken, and the pupil is familiar with the subject, he will learn, although necessarily more slowly than in mounting birds.

I have made this subject a life-study, and have yet to learn a better method, although I think I am familiar with all the various styles of Mammal mounting practised by others. At the risk of being considered egotistical, I will say, that, with *very few* exceptions, I have never seen a Mammal mounted well that was done in any other way.

The preceding will only answer for Mammals smaller than a Mink. In large specimens, when the recent animal is at hand, I arrange the rods as directed, without fastening them to the stand. Then, after making a mould in plaster of the entire body, head, legs, and all, taking care to place the body in some life-like attitude, I place the frame within the mould, and run plaster around it, thus completing the cast. I would advise those who wish to arrive at early perfection in this art, to take a few lessons in plaster-casting. The ears of Mammals should be skinned, and the membrane supplied with thin sheet-lead, fastened to the cast or skull with wire.

Dried skins of large Mammals are mounted upon models made of plaster and clay.

NOTE TO SEC. II. The Dermal Preservative will be found excellent in preserving the skins of mammals, in fact it completely tans them. Take a moderately sized skin, for example that of a fox; after skinning as directed for tanning, keep the skin right side out and rub it well with Preservative, keeping it in a warm room and near a fire; as soon as the skin begins to dry a little, which will occur in a short time, it should be scraped with a blunt knife to remove the inner skin. This peels off in strips; and, as the skin dries, the scraping should be continued, at the same time the skin should be stretched and rubbed, continuing until the whole becomes soft. Skins that have been dried should be soaked in water in which a quantity of Preservative has been dissolved, then treated as above directed.

I now mount mammals in a similar manner as I do birds, with equally good results. Clay and plaster are excellent, but great care must be exercised not to over-fill the animal, for if this be the case the skin in shrinking will invariably pull out the stitches where it is sewed up, or will break in some thin place ; experience will, however, enable one to overcome this difficulty, yet in ordinary mounting I prefer the soft filling. Small mammals are stuffed with cotton, large ones with excelsior, as described in Chapter II., Section III.

CHAPTER III.

Beetles, or Coleoptera.— If Beetles are put into alcohol, they should not remain in it long, but should be either pinned or packed carefully into cotton-wool, when they may be transported in this way.

I have given general directions for collecting Beetles, but will now mention each family in detail.

CICINDELIDÆ, *Tiger Beetles*, are found in dusty roads, on sand-beaches, in rocky pastures,— in fact I have collected them in nearly all places which were devoid of vegetation, from the rocks of Grand Menan, to the barren salt-marshes of Florida. They are very agile insects, but may be caught in nets quite readily. The larvæ live in holes, in sections inhabited by the Beetles. They should be preserved in alcohol.

CARABIDÆ, *Ground Beetles.*— This is a very large family. They are found upon the ground, under stones, chips, and other débris. I have also found some species washed ashore by the sea, often in great numbers. I have taken many of the species of the genius *Lebia* from the flowers of the Golden Rod. All these Beetles prey upon other insects, or feed upon dead animal matter. The larvæ are found in similar situations with the adults.

AMPHIZOIDÆ.— Is a subaquatic family, and is restricted, in its distribution, to California.

DYTISCIDÆ, *Diving-Beetles*, are found very common

97

in the water, and with the larvæ, which are known as Water-Tigers, may be captured with a net.

GYRINIDÆ, *Whirligigs.*— These well-known Beetles are found upon the surface of the water; but the larvæ live beneath it. Both may be taken with the net.

HYDROPHILIDÆ, *Water-Beetles.*— These are small insects, found in ponds and other bodies of fresh water, and, with the larvæ, may be taken with the net.

PLATYPSYLLIDÆ, *Parasitical Beetles.*— The only species which represents this family is found as a parasite, on the American Beaver. I have never found them common, however. Indeed, many Beavers do not have them at all.

SILPHIDÆ, *Carrion Beetles.*— Members of this family are usually large and showy Beetles. They may be captured by exposing the carcass of an animal during the summer. They may be found crawling over it at night, or beneath it in daytime.

PSELAPHIDÆ.— I have captured species of this family flying at twilight. Some are found beneath stones and some in the nests of ants. They are all small insects.

SLAPHYLINIDÆ, *Rose-Beetles.*— These long-bodied singular-looking Beetles are found beneath stones, leaves, bits of wood, etc.

Mr. Henry Hubbard, of Cambridge, informs me that he has captured a great many of this family, and members of other families having similar habits, by gathering leaves and other débris in the woods, and sifting it through a rather coarse sieve over a white cloth. The insects will then fall through. I have found them in large numbers, with species having similar habits, beneath stones, during a dry season in autumn.

Some species are found on flowers, and some in the dry sand of sea-beaches. One, at least, occurs in fungus, and a few species under the excrements of animals.

HISTERIDÆ. — I have found many of these Beetles beneath the excrement of cows, especially in Florida. A great many also occur in carcasses, while one is only found in ants' nests in early spring.

SCAPHIDIIDÆ. — These small insects are found only in fungi.

TRICHOPTERYGIDÆ. — These are the smallest Beetles known, and are found beneath the bark of trees, or in ants' nests.

PHALACRIDÆ. — These small Beetles are found both under bark and on flowers.

NITIDULARIÆ. — These small, flat insects are found beneath the surface of the ground, under bark and stones.

MONOTOMIDÆ are found under bark of trees.

TROGOSITIDÆ occur in grain, and under bark.

COLYDIIDÆ may be found in fungi, or under bark.

DERMESTIDÆ, *Skin-Beetles.* — This is the family which gives so much trouble to collectors.

Dermestes lardarius is a dark-colored Beetle, with greyish buff markings on the base of the wing coverts. The perfect insect gives but little trouble; but the larvæ, which is long, and cylindrical, and covered with red hairs, is particularly destructive. I have known these larvæ to destroy small skins in a few hours.

Anthrenus varius is another museum pest. Both the Beetles and the larvæ attack the skin of the feet, and bills of birds.

The best way to rid skins of both these species is, to saturate them with benzine. This will not injure the feathers in the least. All skins should be kept in insect-proof cases, which I now furnish.

SCARABÆIDÆ, *Horn-Beetles*, are found in decaying wood, animal excrement, on flowers, beneath stones, on sandy beaches, flying at twilight, and in the night. They are mainly large and showy insects. Some are found feeding on the flowing sap of newly-cut trees. The larvæ are found in decayed wood and beneath the surface of the ground.

BUPRESTIDÆ, are found on trees, flowers, and in decayed wood. These are very beautiful Beetles.

ELATERIDÆ, *Snap-Beetles*, occur in decaying wood and beneath stones. The larvæ are called Wire-Worms, and live in the ground.

LAMPYRIDÆ, *Fire-Flies.*— These well-known insect, are found on grass and flowers. The larvæ are called Glow-Worms.

CLERIDÆ, *Flower-Beetles*, are, as the name implies, found on plants and flowers.

TENEBRIONIDÆ, *Meal-Beetles.*— The larvæ are called Meal-Worms, and are, with the Beetles, found about mills, and in grain, flour, etc.

MELOIDÆ, *Blister-Beetles*, are found in the nests of bees, and on flowers, especially on the golden-rod.

CURCULIONIDÆ, *Weevils.*— This is an extensive family. I have captured members in many and various situations. During the early summer months I have taken several from along the seashore, where they were washed ashore. The large Palm-Weevil of the South may be captured in palmetto groves in the evening, or they congregate about the freshly-cut trees, to feed upon the flowing sap. Many species are found in the bark of trees and in fruit. They also occur on flowers and in the stalks of plants.

CERAMBYCIDÆ, *Long-Horneed Beetles.*— I have captured many species in wooded districts during the evening, when they were flying from one piece of woodland to another. They occur in various species of trees, to which the larvæ do much damage by boring into them. Many species are also found on flowers in autumn.

CHRYSOMELIDÆ, *Leaf-Beetles.*— These Beetles are found on leaves and flowers of plants. They may be collected by beating with a stout net. The infamous Potato-Beetle is an example. The larvæ occur on plants.

COCCINELLIDÆ, *Lady-Birds.*— These well-known Beetles are found on plants and trees.

Hemiptera, Bugs.— The well-known Ciccadia belongs to this order. They may be captured with the net. Many species of this order may be taken by beating shrubbery with a stout net.

Orthoptera, Grasshoppers, etc.— I now kill members of this order with the fumes of benzine. They should be placed in a close box which is partly filled with cotton cloth which has been saturated with benzine. In pinning, I now double up the legs of the larger species.

GRYLLIDÆ, *Crickets.*— They may be found under stones, pieces of wood, etc., or in open fields. The Mole-Cricket occurs beneath the ground, in damp localities. Its hiding-place may be detected by its song-notes.

LOCUSTRAIÆ, *Grasshoppers.*— The collector should learn to distinguish members of this family by the song. Some sing only at night, and may thus be captured with a light. They are always found on grass, trees, or plants, and by carefully approaching them, they may be taken in a net or with the hand. Some species which inhabit high trees, like the Katy-did (*Cyrtophyllus concavus*) are difficult to procure. The various species of the genus *Ceuthophilus* and allied genera, called Cave-Crickets, are found under stones, logs, in cellars, and in caves.

ACRYDII, *Locusts*, are found in the grass, on barren rocks and hills, on sandy beaches, on marshes, in meadows, and often in the pine woods, especially in the South. They are best taken with the net. As many of the species have colored wings, the rarer ones are quite easily detected. They are often exceedingly local in their distribution, and members of certain species may be found, year after year, in the same locality.

PHASMIDA, *Walking-Sticks*, may be found on bushes, generally in rocky pastures. Some of the species are found in Florida quite abundantly, on the trunks of trees, or on the grass in the pine-barrens.

MANTIDÆ, *Walking-Leaves*, are found on leaves of plants and shrubs.

BLATTARIÆ, *Cockroaches*, are found in houses, ships, etc.; but some species are found under bark and stones.

FORFICULARIÆ, *Earwigs*.— I have found these insects in great numbers, beneath stones, in southern Florida. Large flights occasionally occur at night, in the north, and as they are attracted by light, they are easily taken.

I do not now recommend placing even the larvæ of *Oithoptera* in glycerine, as it will not preserve the color for any length of time. 　　　　　．

Lepidoptera, Moths and Butterflies.— Especial care should be taken to collect the larvæ of the various species; notes should also be taken as to the food-plants of the larvæ. Mr. Scudder recommends opening the larger species of larvæ behind, and removing the contents of the skin by compression. Then the skin is inflated by means of a straw, and while in this condition it is subjected to the heat of a small oven, beneath which is an alcohol-lamp. The specimen is thus dried slowly, and moulded into a natural form. Then by inserting a copper wire it can be mounted. The smaller larvæ, as well as the pupa, should be placed in alcohol. The latter, if covered with a hard shell may be removed after a few days, and dried.

Casts may be taken of the larger larvæ, and colored as describe dunder "Reptiles and Fishes."

CHAPTER IV.

SECTION I. *Fishes.*— I now make casts of Fishes, proceeding as follows :   Place the Fish side down, on a plate of glass or other smooth surface, and cover it with plaster.   When this is set, remove the Fish from the under side, and varnish the inside of the mould thus formed, and put in cream-plaster; then lay a slab of freshly-cast plaster over the whole, and after the cast has set, chip away the mould.   The cast thus made can be colored to represent life.   Reptiles may be modeled in a similar manner.

Salamanders may be found under logs and stones in damp woods, and also in the water.

104

CHAPTER V.

SECTION I. CRUSTACEA.— Many crabs are found under stones, logs, etc., in the South. Some inhabit holes on the shore or in the woods, and a few climb trees. Some can only be obtained by dredging in deep water, while others always inhabit the shallow margins of bays, etc. Many interesting species are found clinging to seaweed picked up floating in deep water, while others are taken from the shells of living Mollusks. A few species occur in fresh water far from the sea, and at least one has been taken from the caves of Kentucky.

A good way to preserve the smaller species is, to stitch them to cardboard. In transporting Crustaceans, they may be packed in salt; then, afterwards, washed and dried.

COLLECTING MOLLUSKS.

SOLIGINIDÆ AND SOLIGOPSIDÆ *Squid and Cuttlefishes*, may be found floating in the open ocean, or drifted ashore on beaches. They may be taken in nets, or by dredging. They should be preserved in alcohol, or casts taken of them.

PHOLADIDÆ, *Boring-Shells.*— The species of *Teredo* and *Xylotrya* are found in timber that has been beneath the surface of the water for some time. Some of them should be preserved in alcohol, and the shells of others saved, care being taken to keep the parts of individual shells together. Members of *Pholas* and

Zirfœa are to be found burrowing in clay, mud, or rock. Care should be taken in extracting them, as the shells are fragile. If the shell be immersed in hot water for a moment, the animal can be removed with the help of a knife.

SOLENIDÆ, *Razor-Shells*, may be found burrowing in the sand between tide-marks. They may be collected and treated as above, care being taken to tie the valves together, in both cases.

MYADÆ, *Clams*, are found both in mud and sand, usually between tide-marks. They should be treated as other bivalves.

CORBULIDÆ, PANDORIDÆ, ANATINIDÆ, MACTRADÆ, are found either by dredging, or thrown on sandy beaches.

GASTROCHÆNIDÆ are found adhering to marine objects, or embedded in marsh or clay.

TELLINIDÆ may be collected along sandy beaches, between tide-marks.

LUCINIDÆ are inhabitants of deep water, or mudflats which are seldom left dry by the tide.

CYCLADIDÆ are all small, fresh-water bivalves, with quite fragile shells. They should be carefully cleaned and packed in cotton.

CYPRINIDÆ, VENERIDÆ, CARDIADÆ, and ARCADÆ, are all salt-water bivalves, and are generally only to be collected by dredging, often in deep water.

UNIONIDÆ, *Fresh-water Mussels.*— These may be collected in large numbers on river-bars when the water is low, or by dredging in lakes or ponds. I do not recommend boiling them, but they should be exposed to the sun for a short time; then, when dead, they may be cleaned with a knife. Care should be taken to tie the valves together. The outer surface should be oiled slightly.

MYTILIDÆ, *Salt-water Mussels*, may be found on banks left exposed by the tide, or adhering to posts, etc., which stand in the water, or embedded in salt marshes. They may be scalded and cleaned; but care should be taken to preserve the byssus, that is, the ligament by which the shell fastens itself to rocks, etc.

PECTENIDÆ, *Scallops.*— These are found on grassy mud-flats that are not left exposed by the tide. They may be scalded and cleaned.

OSTREIDÆ, *Oysters*, are found on bars, at the mouths of rivers, where the fresh and salt water mingles. They may be scalded and cleaned. Some, however, are found in salt water, clinging to other shells, stones, etc.

TEREBRATULIDÆ.— Found usually in deep water, off the coast.

BULLIDÆ.— These are fragile, univalve shells, found on mud-flats, at low water. They should be boiled, cleaned, and wrapped in cotton.

DORIDIDÆ, TRITONIIDÆ, ÆOLIDIDÆ, ETC.— These are shell-less Mollusks, found adhering to seaweeds. They are to be preserved in alcohol. They form fine objects for the aquarium.

CHITONIDÆ, *Chitons.*— This interesting class of Mollusks which are covered with armadillo-like bands, are found clinging closely to the rocks between tide-marks. The animal should be removed with a knife, and the shells laid flat on a plain surface; then, a board laid over them, to keep them from curling when they dry.

DENTALIDÆ, *Tooth-Shells*, are dredged from great depths. They are cleaned by boiling.

PATELLIDÆ AND CALYPTRÆIDÆ, *Cap-Shells*, cling to rocks, shells, etc. They must be taken unawares from the surface to which they cling, or it will be difficult to remove them, as the animals will contract and adhere very firmly.

ZANTHINIDÆ.— These are thin little shells, highly colored, which float on the open ocean by means of a mass of vesicles. They are quite frequently driven ashore during gales. I have seen vast quantities of them on the Florida Keys.

TROCHIDÆ are found in deep water, but, being small, are frequently washed ashore on sand beaches.

PALUDINIDÆ are fresh-water shells, and are provided with an operculum, which appendage should be preserved and kept with the shell, either by gluing in place after the specimen is cleaned, or by wrapping the shell in paper. The same remarks apply to all operculated shells.

LITTORINIDÆ.— Small shells, found either in salt water or near it, adhering to plants or shrubs. They often occur in great numbers. I have seen the mangroves of the Florida Keys covered with the *Litorina scabia*, and the grass of the salt marshes at Cedar Keys was covered with millions of *Litorina litorea*.

TURRITELLIDÆ.— The so-called Worm-Shells occur in great numbers on the Keys of Florida, growing quite frequently in sponges. Other members of the family are found in deep water.

CERITHIIDÆ are found both in the salt water, often at great depths, and also on salt marshes, near the water. I have seen the ground absolutely covered with some species. Others are found clinging to sea-weeds.

PYRAMIDELLIDÆ.— These are all small shells, and are either found clinging to seaweeds or sheltered by other larger shells.

NATICIDÆ.— Members of this family are quite frequently found on sandy beaches, or in shallow water, near the shore. The animals may be removed by boiling.

TUBRITIDÆ AND COLUMBELLIDÆ. — Found on beaches, and also in mud-flats and in deep water.

PURPURIDÆ. — Found on ocean rocks in muddy creeks, along sandy shores, and in deep water. *Purpura capillus* is very abundant on the rocks, and countless numbers of *Nassa obsoleta* are found in the creeks, from Maine to North Carolina.

MURICIDÆ are found in deep water, and along shores and flats which are not exposed by the tide;·while some are found on mud-flats.

HELICIDÆ, *Land-Shells.* — Land-Shells are found beneath stones, logs, etc., in cellars, and clinging to trees and plants. They are very abundant in some localities. Some are also found in fresh water, especially in springs, where they are sometimes very numerous. They should be carefully cleaned, as much of their beauty and value depends upon this.

ARIONIDÆ. — This is a family of Land-Shells, some of which have little or no shells. They may be found on trees and plants, and should be preserved in alcohol. Others, like the genus *Zonites*, have shells.

PHILOMYCIDÆ are shell-less, and should be preserved in alcohol.

AURICULIDÆ. — Members of this family are found on the land, and also near the salt water, and occasionally in it, as in some species of *Melampus.*

LIMNÆIDÆ. — These genera are all represented by species which live in fresh water, often in rivers and lakes.

SPIRULIDÆ are found in the open ocean, and occasionally drift on shore.

All shells should be carefully labeled with date and

locality in which they are collected. Notes as to the relative abundance, etc., should be made.

Many species of the deep-water species may be obtained from the stomachs of such fishes as the Cod. Others may be found in the gizzards of Ducks.

Small shells which cannot be cleaned should be placed in alcohol, and allowed to lie for at least twenty-four hours, then taken out and dried in the shade.

Some shells, like the thin-shelled Unioes, are liable to crack when dry; if the fresh shell be dipped into a solution of chloride of calcium, this will be prevented.

CORALS.

GORGONIAS, *Sea-Fans*, *Sea-Pens*, frequently grow in comparatively shoal water. I have often seen them left exposed by the falling tide; at such time, they may be gathered in large quantities, for they are almost always abundant. They may be dried carefully in the shade; then they will not lose their brilliant colors.

MILLEPORAS *and other branching Corals.*— Some species are found on reefs that are exposed at low tide, but some must be obtained by dredging. A good machine for collecting is made in the following manner: Procure a bar of iron five feet long, one inch thick, and three inches wide; have holes one inch in diameter drilled, one inch apart, for the entire length. Next, have two eye-bolts fastened in near each end. Now, pass ropes, one inch in diameter and five feet in length, through the holes, taking care to knot them at the ends, to prevent their going entirely through. Then, unravel the ropes and fasten stout lines or chains to the eye-bolts. This is thrown over-board, and dredged over the bottom, when the Coral will become entangled in the trailing-ropes, and brought up.

ASTRÆCEA, FUNGACEA, ETC., *Brain and Mushroom Coral.*—This form of Coral is, perhaps, the hardest to collect. They may be procured by expert divers, who break them loose with hammer and chisel, or they may be broken off with a sponging-hook, and thus brought to the surface.

PRESERVING CORAL, BLEACHING, ETC.

CORAL that is to be preserved with the natural color should be carefully dried in the shade ; but all Corals may be bleached nicely, by dipping in salt water once a day, and exposing to the sun. They should be first killed by exposing them for a few hours to the heat of the sun; then allowed to lie for twenty-four hours in water, when the gelatinous animals will be dissolved, and will run out of the cells. Then they should be rinsed and exposed to the sun.

ACTINÆ, *Sea-Anemones,* may be looked for at low tide on rocks, or found by dredging (as the deep-water species adhere to shells, stones, etc.). They may be removed from a smooth surface by carefully sliding along until some is introduced beneath the sucking disk, when they will become loosened; or they may be taken off by means of the blunt edge of a spoon or some similar instrument. These objects, which are so beautiful while living, possessing colors which vie with the flowers, are exceedingly difficult to preserve. They change very much in alcohol ; but Prof. A. Hyatt informs me that he has found picric acid the best medium by which the colors may be kept.

HYDROIDS AND BRYOZOA.—I have found quantities of these in shoal water. They are very delicate, and should be carefully handled.

Star-Fishes.— I have at times, found the beaches covered with some of the common species, but the best way to find them is to visit the rocks where they occur at low tide; in favorable localities they may be collected by thousands. The larger species occur on flats, while some are only to be obtained by dredging. Some species shed their arms quite easily; these should either be kept in salt water, or thrown at once into alcohol. Indeed, all species are best kept in water until they assume a form in which they are to be dried; then they should be instantly plunged into strong alcohol, where they must remain for at least twenty-four hours. Then they may be removed and dried in the shade. When quite dry, I have found it advantageous to dip them into a solution of hot paraffine, as this prevents their disintegrating, as they are subject to this trouble.

ECHINODERMS, *Sea-Urchins*, cling to rocks, and may be found at low water or dredged in deep water. In killing them, observe the same precaution as practised in Star-fishes, and then treat them in the same way.

Sponges.— Those who are only familiar with the dried and prepared skeletons of the sponges of commerce would be surprised to see one of these animals in its native state; indeed, I scarcely think the uninitiated collector would recognize the accustomed sponge in the gelatinous mass which grows on the banks of the Florida Keys. Sponges are of various forms, some of which are very beautiful. They are also of varying colors. Some are exceedingly fragile, while others are, when dry, as hard as wood. Some are largely made up of siliceous matter; indeed, long, needle-like crystals of silica are to be found in greater or less quantities in almost all species.

They grow in various situations; some few occur in fresh water, but they mainly inhabit the sea. I have found them encrusting rocks, pieces of shells, growing on submerged posts, and one or two species I have taken from the backs of the more sluggish species of Crabs. A few float, or rather roll, about on the bottom; these usually occur in secluded bays or sounds, where there is but little sea. The greater portion are found attached to the bottom, often in great depths of water; but many are found growing on banks, in from three to four fathoms. I have also collected a great many in exceedingly shoal water, near the shore. I have seen thousands of the deep-water species, washed ashore by the actions of the waves, during hurricanes.

The sponges which occur in shoal water may be removed by help of a knife; but the deep-water species are gathered with a sponge-hook, which is simply a two-pronged fork, bent into a hook. The spongers of the Bahama Islands and the Florida Keys use what is called a water-glass, to discover the sponges on the bottom. This is a cylinder of wood or metal, of about ten inches in diameter, having a glass bottom. This is placed in the water, bottom down, and thus the operator can see the sponges quite easily through the glass, even if there is a ripple on the water, which would otherwise prevent his seeing to any depth.

Sponges should be dried in the shade, without exposing to the sun, care being taken that they do not absorb dampness from the atmosphere, as then they are apt to decay.

The skeletons are procured in the following way: The sponges are first exposed for a day or two to the influence of the sun, then are thrown into what is known as a "sponge crawl," which is simply a large pen with slatted sides, placed in some swift tide-way. They are kept here until thoroughly macerated, which process occupies about a month. They are then stirred about, and beaten until all the animal matter is removed, when they are dried. Many of the sponges of commerce are bleached with chloride of lime, which, although it improves their looks, is at the expense of the durability of the tissues.

SECTION II. *Preparing Skeletons.*—I do not now recommend using chloride of lime. The bones should remain under water, in some glass or wooden vessel, until every particle of fleshy matter is dissolved. If the bones are oily after they are bleached, they should be once more immersed in water, and kept there until clean. Skeletons of large Mammals often require more than a year to macerate them properly. I now remove all horny portions, even of the feet and bill. In articulating small bones, I now make use of a very strong cement, which is advertised in my catalogue of taxidermists' supplies.

CHAPTER VI.

COLLECTING AND PRESERVING EGGS.

Too much care cannot be taken to clean eggs. The contents must be thoroughly removed, to accomplish this, and all eggs should be rinsed. Eggs of all species should be kept in sets; and, where it is possible, the nest should be collected with them. In packing eggs for transportation, wrap each one separately in cotton, and place in a box lined with the same material. Capt. Chas. Bendike, who is the most careful egg-collector that I ever met, and his beautifully-prepared and extensive collections bear ample testimony to this statement, packs eggs in the following manner: The box is first lined with cotton-batting; then the eggs are placed, side by side, in partitions made of the same material, which is cut in strips for this purpose; then a layer of cotton is placed over them, more eggs packed, and so on until the box is filled. 115

www.ingramcontent.com/pod-product-compliance
Lightning Source LLC
Chambersburg PA
CBHW021936190326
41519CB00009B/1033